W9-DBK-254

THE HAIR
OF THE DOG

Also by Karl Sabbagh

The Living Body

Skyscraper

Magic or Medicine? (with Rob Buckman)

Twenty-first-century Jet

A Rum Affair

Power into Art

Dr Riemann's Zeros

Palestine: A Personal History

Your Case is Hopeless

Remembering Our Childhood

THE HAIR OF THE DOG

AND OTHER SCIENTIFIC SURPRISES

KARL SABBAGH

JOHN MURRAY

First published in Great Britain in 2009 by John Murray (Publishers)
An Hachette UK Company

5

© Karl Sabbagh 2009

The right of Karl Sabbagh to be identified as the Author of the Work
has been asserted by him in accordance with the Copyright, Designs
and Patents Act 1988.

All rights reserved. Apart from any use permitted under UK copyright
law no part of this publication may be reproduced, stored in a retrieval
system, or transmitted, in any form or by any means without the prior
written permission of the publisher, nor be otherwise circulated in
any form of binding or cover other than that in which it is published
and without a similar condition being imposed on the subsequent
purchaser.

A CIP catalogue record for this title is available from the British Library

ISBN 978-1-84854-088-0

Typeset in Scala by Servis Filmsetting Ltd, Stockport, Cheshire

Printed and bound by Clays Ltd, St Ives plc

John Murray policy is to use papers that are natural, renewable and
recyclable products and made from wood grown in sustainable forests.
The logging and manufacturing processes are expected to conform to
the environmental regulations of the country of origin.

John Murray (Publishers)
338 Euston Road
London NW1 3BH

www.johnmurray.co.uk

CONTENTS

INTRODUCTION

Over the many years in which I have taken an interest in science, a ragbag of facts, anecdotes and oddities have stuck in my mind, and some of them form the substance of the short essays in this book.

Here you will read about an aircraft carrier made of ice and sawdust, a nuclear blast detector using ESP, a scientist who was strangled by his own invention, a philosopher who believes that if there were more women scientists science itself would be fundamentally different, a poor black woman whose cancer cells are now in every laboratory in the world, and a two-billion-year-old nuclear reactor. You will find the answers to questions you might never have asked, such as: Why is the sky dark at night? Why doesn't the Earth move when we move our eyes? Is your brain really necessary? Can money bring happiness? Can the blind see? Does 1 + 1 really equal 2?

But these essays are more than amusing or intriguing curiosities. Many of them help us understand how astronomy or maths or biology or physics actually work, and the methods scientists use to verify their hypotheses.

For example, the answer to 'Why is the sky dark at night?', first asked hundreds of years ago, depends on knowledge acquired in the twentieth century about the expansion of the universe. Asking 'Can the blind see?' takes us into a pioneering area of neurophysiology. And even the existence, regrettable, perhaps, of a molecule called 'arsole' helps us understand the systematic way in which scientists name new biological molecules.

HEAVEN AND EARTH

IS THIS A RECORD . . .?

Here's a prediction: in 40,000 years' time, in a planetary system around a nearby star, a group of aliens will gather around a gramophone, one of those devices we dimly remember from the 1970s. They will lift the pickup arm and place the stylus gently on a disc rotating at $33^{1}/_{3}$ rpm. They will then hear – if they have ears to hear – a greeting in the Akkadian language used in Mesopotamia about six thousand years ago.

As predictions go, this sounds fairly dotty. But it is based on sound reasoning. In 1977, NASA sent two spacecraft, Voyagers 1 and 2, off into space. Their mission was mainly to gather data, including photographs, about Jupiter and Saturn and their moons, and then to travel on out into space, across the vast emptiness between our solar system and the nearest stars.

Each Voyager was the size of a small car, and carried what was described at the time as a 'phonograph record', with the intention, or at least the hope, that one day in the distant future some extraterrestrial civilization would discover one of the Voyagers, detach the interesting-looking disc and 'play' it. If they did so, they would find an hour's worth of sound, including greetings in many Earth languages and various earthly sounds, such as a bus, a tractor and a horse. The disc also contains 115 still photos, including one of a woman eating an ice-cream cone and another of a man eating a piece of pizza.

But in 1977 the latest technology for recording and playing back sound was the long-playing record. The first patent for what became the CD wasn't filed by Sony until 1980. So what are aliens to do in 40,000 years' time, the soonest the Voyagers could possibly be intercepted by inhabitants of other planets round other stars? Faced with a disc with a fine groove spiralling from the outer edge to the centre, how are they to know that its tiny variations conceal information about our far-off planet and its preoccupations in 1977?

To give the aliens a sporting chance, on the cover of the record NASA printed a picture of a gramophone stylus seen from above and from the side. Since no written instructions would have been comprehensible, the NASA scientists tried to convey in diagrams such information as the way the stylus worked, the need to rotate the disc, what the period of rotation should be, and how to translate the squiggles detected by the stylus into sounds or pictures. And so that the aliens knew exactly where the spacecraft had come from, NASA engraved a map, showing the location of the Solar System in relation to fourteen pulsars, stars with a particular, very precise period of rotation, which would be likely to be known to the aliens.

The whole project is a triumph of hope over expectation. It's difficult enough to imagine how *humans* of the distant future might interpret such a disc. The pace of technical change is accelerating so fast that the world of AD 42,000 will be unimaginably different from the world of 1977. People might understand NASA's stylus in the way it's intended, or they might think it's a drawing of a train going round a circular track, or they might just pick the record up and throw it to each other like a frisbee (a mid-twentieth-century leisure device, now almost obsolete). At least those humans will probably still have eyes and still be able to interpret diagrams. But who knows what sense organs will be possessed by intelligent lifeforms elsewhere in the universe to help them interpret the disc? What engineering technology will they have to manufacture the stylus, the turntable, the speed con- troller, the loudspeaker? Will they even have ears to hear or eyes to see, once they put the record on their celestial gramophone?

It's easy to ridicule the idea of sending a 'how to' kit to make a gramophone into space, at a time when it is becoming the norm for sounds and images to be stored on hard disks with a capacity that increases tenfold every year or two, and with iPods capable of storing *eighty days'* worth of sound rather than the one hour on the Voyager disc. In fact, such new technology is at the heart of a more recent space project, called KEO, which is planned to launch in 2010 or 2011, carrying messages from as many inhabitants of Earth who choose to participate, encoded in glass radiation-resistant DVDs. Needless to say, the spacecraft will carry instructions for the future recipients on how to build a DVD player. And also needless to say, it can only be a few years before this too will seem as out of date a way of recording and reading data as the Voyager gramophone is today. But unlike the few hours' capacity on the Voyager disc, the KEO DVDs will have enough space and compression to allow every citizen on Earth today to send a message up to four pages long.

Not everyone thinks this kind of reaching out to other life- forms is a good idea. At the time of Voyager, Professor Sir Martin Ryle, the leading British radio astronomer, protested to NASA that, by revealing our presence and exact location, we would be making it possible, if the aliens are malevolent or even just hungry, for them to invade our planet and put an end to the human race, assuming there still is one. But the time for such worries was past even in 1977. Any extraterrestrial civilization worth its kryptonite will be fully aware of the details of life on Earth by watching the programmes of David Attenborough, along with the other outpourings of terrestrial television. Whether, on balance, they would conclude that Earth was a place worth avoiding rather than visiting is at least arguable.

AN EXPLODED VIEW

In 1991 the British artist Cornelia Parker created an installation called *Cold Dark Matter: An Exploded View*. It was produced by filling a garden shed with objects gathered from sheds belonging to herself and her friends and arranging for the British army to blow it up. Parker then collected the pieces and created a work that represented a moment after the explosion as the shed and its contents flew apart.

Scientists are fond of 'thought experiments', where new ideas can arise as a result of imagining an experiment that would be very difficult, or even impossible, to do. One thought experiment with Parker's shed would be to imagine the pieces flying apart and then to reverse the direction of travel of each piece so that they all converged on their original positions, reconstituting the shed and its contents. It is that sort of imaginary process that cosmologists have carried out with the universe, with some very surprising results.

The universe is expanding, with stars and galaxies flying

6

away from each other like the pieces of Cornelia Parker's shed. Scientists have arrived at a picture of the history of the universe by imagining its contents retracing their steps. Using their knowledge of the masses and velocities involved, and a lot of physics, they have traced the history back over a period of 12–15 billion years to a point in time and space where our universe could be said to have begun, with the event known as the Big Bang. For much of that time, the universe would have behaved like Parker's shed if run in reverse, and any of us, even the non-scientists, can imagine the stars and galaxies (and, indeed, some material called cold dark matter) converging at speed on some central point. But going back past about 300,000 years after the Big Bang, the whole picture changed. By applying the laws of atomic physics at high temperatures and pressures, a picture emerges of the first few moments in the life of the universe which is as strange and unimaginable as any painting by Dali or Magritte. It is as bizarre a picture as if Parker's shed had continued contracting after it reached its initial state, and as it got smaller and smaller it changed shape, its wood and metal and cloth components turned to cheese and neon and diamond and eau de cologne, and eventually shrank into a tiny dot the size of an atom with the temperature of the heart of the sun.

'If the history of physics has taught us anything,' wrote two physicists recently, 'it is that the true nature of the world may lie beyond our ability to visualize directly.' This is all very well, but humans are creatures whose first instinct is to visualize when trying to understand complex scientific explanations. Nevertheless, as science becomes more theoretical and less dependent on direct observation, it is more difficult for non-scientists to grasp what scientists are on about.

The theory that purports to describe the early years, days

and seconds of the universe is firmly based. It successfully explains observations that puzzled astronomers and cosmologists in the past, and has led to new experiments using satellites based on predictions that were subsequently fulfilled. In the past, people who believed that the Earth was round bolstered their theory with the well-known observation that ships setting out from land seemed to disappear, almost as if they were sinking, at a point in their voyage where they should still be visible. This 'round Earth' theory also predicted that a ship would be able to sail continuously from east to west and arrive at its starting point. The predictions and observations confirmed the theory of a round Earth that everyone now believes. A similar combination of observations and predictions underlies scientists' understanding of the birth of the universe.

Here's what they believe happened:

During the first *second*, the shortest period of time any of us can imagine realistically, the universe expanded from an object that was infinitely hot and with a radius of zero to a sphere with a radius of four light years (about 40 million million kilometres). It also cooled down to 10,000 million degrees. (The temperature at the heart of the sun is only 15 million degrees.) During that first second, a series of events took place one after the other which I list below, not in the hope that you will understand what they mean (I don't) but to show the nature of the scientific imagination at work, unshackled by the need to relate the ideas to the real world:

- 'The quantum wavelength of the universe was larger than the size of the universe itself.'
- 'At the Planck time, symmetry breaks. Gravity becomes a distinct force. We have entered the GUT

era. Here we have the quantum limit of classical general relativity.'

- 'Limit of perturbative interaction – thermalization of universe.'
- 'Grand unification and spontaneous symmetry breaking.'
- 'The universe enters a state called a "false vacuum".'
- 'Electroweak era begins.'
- 'End of Electroweak era. The Electroweak force breaks down into two distinct components, the weak nuclear force and the electromagnetic force.'

(These last two events are worth noting in passing. The Electroweak 'era' lasted from 1/100,000,000,000,000,000,000, 000,000,000,000,000th of a second after the beginning of the universe to 1/100,000,000,000th of a second, surely the shortest 'era' in history.) By two ten-millionths of a second after the Big Bang, the universe was the size of the solar system and had cooled to 1,000,000,000,0000 degrees.

I could go on and describe a similarly mind-boggling series of events that took place between the first second and the first minute, the first minute and the first hour and, in fact, right up to about a billion years after the Big Bang, when the universe began to look a bit like the place we live in today.

What is significant about all this is not that non-physicists are expected to understand and remember these stages, but that mathematics and physics demand a parallel type of understanding to that which operates in most of us in our everyday lives. So often we try to understand what scientists say by getting them to put it into 'ordinary language', but if ordinary language could convey scientific truth, scientists would be using it anyway. 'Ordinary language' says that the universe started as an infinitely small point of infinite temperature, which just sounds confusing – or nonsensical. The best attempt I've seen to convey what this actually means suggests that *all* of space was filled with energy at the beginning and that the expansion from an infinitely small, infinitely dense, infinitely hot point is how our *observable* universe began, from just one point in that mass of energy, and that we have no way of knowing or inferring what happened to the rest, although it could be that every point in the entire universe expanded in the same way, leading to other universes beyond our grasp.

Ultimately, the only terms in which anyone can really understand what happened at the beginning of the universe are mathematical terms, and to understand them we have to learn another language, mathematics. Or we can just use an interpreter, a scientist, which is what most of us are happy to do.

COSMIC COINCIDENCE

A total solar eclipse is one of the most beautiful sights in the sky. Most people who see it never forget the moment when the dark disc of the moon fits *exactly* over the bright disc of the sun, and for a minute or two the gaseous envelope that surrounds the sun becomes visible, temporarily freed from the dominance of the overwhelmingly brighter solar surface. This spectacle occurs once or twice a year somewhere on Earth, although people in any one spot will have to wait nearly four hundred years to see it. Nevertheless, it was known to ancient cultures around the world, from Greece to Central America, and they built it into their pictures of the cosmos as a rare, dramatic and significant event.

And yet, it all comes down to a cosmic coincidence. If the moon were a little smaller or farther away, no one would have been aware of the beauties of an eclipse. (If the sun were larger or nearer, we would be unlikely to be here to be aware of anything.) The exact superimposition of the moon over the sun comes about only because of the figure 400. The sun's diameter is 400 times larger than that of the moon, but it is also 400 times farther away, which means that their apparent diameters are the same. But there is no physical or astronomical reason why this should be so. The way the moon formed, probably as the result of a giant impact between the Earth and another celestial body, has nothing to do with the distance or size of the sun, and the resulting size of the moon after that collision was determined by gravity and the movement of the fragments of the collision in the region local to the Earth.

One astronomer, Guillermo Gonzalez, a believer in the Anthropic Principle [🐕 28], has stuck his neck out and said that he thinks there is a connection between intelligent life on Earth and the phenomenon of a total eclipse, pointing out that the moon has gradually moved farther from the Earth over millennia, reaching the point at which total eclipses became possible at the same time as intelligent life. But then he makes the even more astonishing inference that in other solar systems with other planets, intelligent life will arise only on planets that experience total solar eclipses.

This coincidence of size, by the way, will not last for ever. People who live on Earth 200 million years from now will no longer see the full glory of a total eclipse because the moon will have moved too far away. The best they can hope for is an annular eclipse, which looks like a bright circle of solar light around a dark smaller core. This type of eclipse is the best that Martians can experience as well. One of their moons, Phobos, is about three-quarters the apparent size of the sun seen from Mars.

Here's one final interesting point about our judgement of apparent size: asked to guess how much of the width of the Sun would be obscured by holding a thumb up in front of it at arm's length, most people hazard a guess that the sun's diameter is about the width of a thumb. In fact, it's less than a quarter of the width of a thumbnail, as you can easily verify by experiment. (And if you have a small thumb, don't worry that you'll get a different result from your large-handed friend, because you'll also have a shorter arm than him.)

DROPPING IN ON THE BIG BANG

In the days of analogue TV – fast disappearing as the digital changeover takes hold – older television sets did not have remote controls or even channel buttons, but a rotating knob which you turned through various UHF frequencies until you found the channel you were looking for. If you stopped tuning before you found a channel, you'd see a screen full of 'snow' – white dots coming and going randomly. Some of that 'snow' is a sign of the Big Bang [🐕 7], although nobody was aware of its significance until the 1960s. And even then, a heap of bird droppings got in the way of one of the most significant twentieth-century discoveries in science.

When two physicists working at Bell Labs in the United States tried to find uses for a defunct communications antenna, they noticed some annoying interference which occurred wherever they pointed the antenna in the sky. They'd been hoping to use the antenna to detect radio waves from certain stars, but clearly they would have to remove all sources of interference before they could measure the stellar radiation, which would be very low-level anyway.

The scientists, Arno Penzias and Robert Wilson, climbed up on the roof where the antenna was kept and found what they thought was the problem – layers of droppings from pigeons that had turned the antenna into a home. The droppings were warm, and since the kind of radiation causing the interference was an indication of temperature, they realized they had probably found the source.

After thoroughly hosing down the antenna and scattering pesticides, the scientists went back to their measurements. Sure enough, there had been an effect from the bird droppings, and the level of interference was now lower. But it was still there.

At the same time as Penzias and Wilson were wrestling with their little technical problem, another physicist, Robert Dicke, had been working on what was then a new idea about the origin of the universe, the Big Bang theory. One consequence of this theory was that the universe began in an intense burst of very hot radiation, and spread in all directions over the succeeding billions of years, becoming cooler at the same time. If this theory is correct, the universe today, after 15 billion years, would be bathed in much cooler radiation, at exactly the frequency of the interference found by Penzias and Wilson – after they had cleaned their antenna.

When Dicke heard about their problems, the three scientists got together and it became clear that the Bell Labs antenna had detected the first hard evidence for the Big Bang theory, and in

1978 Penzias and Wilson (but not Dicke) shared the Nobel Prize for physics. Although Wilson had supported a rival theory for the origin of the universe, the Steady State theory, he had to accept that the combination of Dicke's reasoning and his and Penzias's own data supported the Big Bang.

There was another odd twist to the story. It turned out that a paper in a Russian journal in 1964, the year before Penzias and Wilson recognized the source of the interference, had predicted that if the Big Bang theory was correct it would generate microwave radiation, and that the best available instrument for detecting it, because of its size and shape, would be the Bell Lab antenna. Unfortunately, as Penzias said in his Nobel lecture, 'it appear[ed] to have escaped the notice of the other workers in this field', including, presumably, himself.

HENRIETTA'S MILESTONE

On a very clear moonless night, the most prominent feature in the sky is a diffuse band of light that stretches from one side of the horizon to the other. In addition to this band, scanning the sky with the naked eye reveals stars of different brightnesses, and small fuzzy areas which show up in a telescope as galaxies and nebulae. The band of light, which we call the Milky Way, is actually 'our galaxy' – a whirlpool of stars shaped like a disc which we see edge-on because our sun is embedded in it. Until a hundred years ago, astronomers thought that our galaxy constituted the entire universe, and that the stars and fuzzy areas we could see were all part of that galaxy. It was a discovery by one woman, Henrietta Leavitt, which helped to overturn the prevailing picture and revealed a vast and expanding universe in which our galaxy was a tiny component in a much bigger and more complex system, containing millions of 'Milky Ways' whose existence had been undetected by the existing technology.

The essence of Leavitt's discovery was a method of measuring the distances of stars and galaxies. It was almost like discovering in each distant astronomical object a label that said 'I am 56 million light years away', or whatever the actual distance was.

If all the stars, including our sun, were of equal brightness, we would have no trouble measuring distances in the universe. The dimmer a star looked, the farther away it must be. With powerful telescopes we could measure the apparent brightness of stars far too dim to be seen with the naked eye, and with mathematics we could work out their distance. A law of physics called the Inverse Square Law says that with two stars of equal brightness, if one is twice as far away as the other it will look a quarter as bright. If it was three times the distance it would look a ninth as bright, and so on ($1/2^2 = 1/4$; $1/3^2 = 1/9$). So finding the relative distance of stars would mean starting with a star whose distance we know, the sun maybe, and comparing the brightnesses of all the other stars we could see.

But stars are not equally bright. They are of different ages and at different stages of evolution and as a result have different temperatures. Early modern astronomers tried to arrange them in some kind of sequence, from A to S, hottest to coldest, but later discoveries muddled up the order, so they now run: O, B, A, F, G, K, M, R, N, S. I include this piece of information only so that I can tell you that astronomers remember the temperature categories of stars with the sentence 'Oh, be a fine girl, kiss me right now, sweetie'. (Some astronomers in the 1970s used 'On Bad Afternoons, Fermented Grapes Keep Mrs Richard Nixon Smiling'.)

Because stars are not equally bright, we need a way to tell how bright each one actually is before we can determine its distance. Let's say we know by other methods that a star, Alpha, is four light years away. Suppose we know that another star, Beta, is in absolute terms half as bright as Alpha. In other words, if the stars were at the *same* distance, Beta would look half as bright as Alpha. Suppose now that Beta actually looks one sixteenth as bright as Alpha. By the Inverse Square Law, if Beta were as bright as Alpha, it would be four times farther away. But because it's half as bright, we know that it is actually nearer, at twice the distance of Alpha. The point I am making is that if we know how bright a star really is, we can tell how far away it is.

In 1904, Henrietta Leavitt was working for 30 cents an hour at the Harvard College Observatory. She worked in the photometry department, checking hundreds of photographic plates to measure the brightness of stars. Her job required a good eye, an accurate memory and a high boredom threshold.

Although most stars have a fixed magnitude – the apparent brightness as seen from Earth – there are many with fluctuating brightnesses, called variable stars. Because Henrietta Leavitt had a good visual memory, she could look at a photographic plate taken one night and notice whether one star was a different magnitude from when it had appeared on a plate a week before. In this way she discovered more than two thousand variable stars, about half the known total at the time. This was an impressive achievement, but her crowning discovery related to one class of these variable stars, which became known as Cepheid variables, because they had a similar pattern of variability to a star in the constellation of Cepheus.

She noticed that these stars varied in a very regular way – the greater their absolute brightnesses the longer their period of oscillation. So a Cepheid variable that was about eight hundred times as bright as the sun went from bright to less bright and back again over a period of three days (its 'period'), while a Cepheid that was ten thousand times as bright as the sun varied over a period of thirty days. What this meant was that by measuring the *period* of a Cepheid variable, astronomers could measure its absolute *brightness*, and once they knew that, using the method described above they could tell how far away it was.

Leavitt's discovery eventually led to a huge expansion of our estimates of the size of the universe. Using powerful telescopes, astronomers discovered some stars with this pattern of variation in nebulae and galaxies that had been assumed to be part of the Milky Way. But by calculating the stars' absolute brightness from their periods of variation, they found that the stars could not be in our galaxy, since if they were they would look brighter than they did. The fact that they looked much dimmer and yet had a large absolute brightness meant only one thing – they were much farther away than the stars in our galaxy.

Our galaxy is 100,000 light years across – it takes light 100,000 years to travel from one end to the other [🐕 35] – and it turned out that the first galaxy whose distance was calculated from its Cepheid variables was 2.5 million light years away, enlarging the size of the known universe twenty-five times.

For someone who made such a major contribution to astronomical science, Henrietta Leavitt is considered by some people to have been denied the credit that she deserved. Astronomy was certainly a male profession at the time she worked at Harvard, and because she was not a professional astronomer (and perhaps because she was a woman) she was never allowed to use a telescope professionally, although she was passionate about astronomy. Another female astronomer, Cecilia Payne-Gaposchkin, said that preventing Leavitt using a telescope was 'a harsh decision, which condemned a brilliant scientist to uncongenial work, and probably set back the study of variable stars for several decades'.

After Leavitt's death, when her contribution was more widely recognized, her name was given to a 65-kilometre-wide lunar crater – on the back side of the moon.

ALL IN A SPIN

Sometimes, if you read too much science it can make your head spin. The farther reaches of physics – astronomy and cosmology in particular – deal with objects that are presented by scientists in a matter-of-fact way but which challenge one's credulity if you stop to think about them for too long.

I'm not talking about the true exotica of astronomy – black holes, wormholes, multiverses, inflation, and so on. Even more 'ordinary' objects in the sky, described in terms that don't require any specialized knowledge to understand, sometimes tax one's belief.

Take the Crab Nebula. This is a fuzzy patch, visible in a small telescope, which is the remains of a star that was observed to explode in AD 1054. Today (or rather 6,000 years ago, since that's how long light takes to travel from the Crab Nebula to Earth) much of the star's material is seen to be flying away from the centre at a speed of about 1,500 kilometres per second, and it has now reached a diameter of about eleven light years (100 million million kilometres). But that's not the incredible bit. Large sizes and distances I can manage. It's what's at the heart of the Crab Nebula that is the problem.

All that remains of the original star that exploded is what's called a neutron star, in the centre of the nebula. This is twelve miles (about 20 kilometres) in diameter. That's the diameter of London out to the suburbs, or the length of Manhattan from the tip to the Bronx.

This star, a sphere, has a mass of about twice the mass of the sun. The sun has a mass of about 330,000 Earths. So we have to think about a sphere the size of London or Manhattan but weighing over 300,000 times the weight of the entire Earth.

So far, so astounding. But now it turns out that this sphere is spinning, quite fast. In fact, the neutron star at the heart of the Crab Nebula completes thirty full rotations every second. So if you were standing, say, in Richmond Park, where the surface of the neutron star, transplanted to London, rotated past you it would be travelling at a speed of about *four million miles an hour*. (You might like to treat it like a Fermi problem [🐴 272] and work it out for yourself.) This is some way from the speed of light, about 670 million mph, but is still a pretty impressive rotational speed for an object as massive as 300,000 Earths.

The fact is, of course, that if you were standing anywhere near the surface of this neutron star, and you could withstand the heat and X-rays pouring out of it, you would be sucked by gravity into its heart with such force that all the molecules of your body would be torn apart into their constituent atoms. Still, at least you wouldn't feel anything, since this would all happen before a pain signal had travelled more than a fraction of a centimetre along your nerve fibres.

'WE ARE STARDUST. . .'

The American singer Joni Mitchell wrote a song called 'Woodstock', which contained the words 'We are stardust', a lyric that might be taken as no more than a beguiling metaphor, as in 'We'll build a stairway to paradise' or 'Catch a falling star', if it wasn't for the fact that she also sang about 'billion-year-old carbon', suggesting that she knows whereof she sings.

The chemical elements that make up our bodies, including the iron in our blood and the calcium in our bones, originated, with one exception, in the turbulent body of a star. The exception is hydrogen, the most abundant element in the universe, which formed the starter material for all stars and still provides the fuel for the nearest star, our sun. Hydrogen is a simple element and it can be visualized as having a single particle, a proton, as its nucleus and another, an electron, 'orbiting' round it, although this model is now considered oversimplified. It is the building block of the other chemical elements, which have more nuclear particles and more electrons.

In a sequence of events during the life cycle of some stars, hydrogen atoms under pressure combine to produce heavier elements like helium, and the helium under further high pressure and temperature forms carbon and oxygen. By this stage the original atoms consisting of one proton and one electron have combined to form larger atoms, some with six protons and electrons (carbon) and others with eight of each (oxygen). As the star becomes denser and denser the force of gravity crushes these atoms together to produce yet heavier ones, until the process stops with iron. With iron atoms, made up of twenty-six protons and electrons, the process of crushing together to make even heavier atoms stops, and the star's core acquires more iron and becomes heavier and heavier until the star collapses under its own weight.

The process of getting from hydrogen gas to a solid iron core takes about ten million years, but all that hard work is undone in the space of less than a second as the star collapses. Then a shock wave travels out from the centre of the star and blows off the outer layers, which contain a range of elements that were intermediaries on the way from hydrogen to iron – silicon, oxygen and carbon. This explosion results in a huge emission of light and other energy over a few days, and can be observed as what's called a supernova [🐕 188], when a star seen from the Earth increases many times in brightness before lapsing back to a dimmer state.

Those heavier elements, forged in one star, now spread out into space, eventually to form other stars like our sun, which was formed out of a cloud of matter attracted by gravity towards a central point. Some of the elements in that cloud condensed into planets, including the Earth, and thus all the heavier elements that started out in the outer shell of a star came to their final resting place, on the Earth's surface and in its atmosphere. From there it was a short step for a small proportion of that 'stardust' to become incorporated in our bodies, in the following proportions: oxygen (65 per cent), carbon (18 per cent), nitrogen (3 per cent), calcium (1.5 per cent), phosphorus (1.0 per cent), potassium (0.35 per cent), sulphur (0.25 per cent), sodium (0.15 per cent), magnesium (0.05 per cent) and copper, zinc, selenium, molybdenum, fluorine, chlorine, iodine, manganese, cobalt, iron (0.70 per cent).

In fact, it's unlikely that the hydrogen in our bodies – 10 per cent of the mass – came from the explosion of a distant star since there's much more available as gas between the stars, and so it might be better to change the words in Joni Mitchell's song to 'We are 90 per cent stardust. . .'

COMBING THE UNIVERSE

The Doppler effect [🐕 168] is an essential tool in astronomy. It was the source of our understanding of the fact that the universe is expanding, because the light from certain stars and galaxies is redder than we would expect, owing to what is called the 'red shift' in its spectrum as it moves away. The red shift with light is similar to the drop in pitch of sound waves as a source of sound moves away from the hearer.

Until recently, the sort of expansion speeds that could be measured were of the order of 30,000 kilometres per second. When a galaxy is moving at that sort of speed away from us, the change in the colour of the light emitted is large enough to be easily measurable.

The way the change is measured is as follows:

Most of us are familiar with the rainbow colours of the spectrum of white light. If you shine light through a prism on to a sheet of white paper, you get a spread – or spectrum – of colours from red through yellow and green to blue. All of these colours are normally mixed up in one beam of white light and the triangular glass shape of the prism spreads them out. Now, if that source of white light was moving very fast away from the prism, the spectrum would be shifted. The violet light that was at one end of the spectrum would now look blue, the blue light would look green and the green light would look yellow. Everything would be 'shifted' towards the red end of the spectrum. As it happens, with white light from, say, the sun, a frequency or 'colour' of light that had been invisible beyond the violet end (ultraviolet) would now be violet, and so in fact the spectrum wouldn't look any different to the naked eye. But physicists have found a way to detect this shift, using what are called emission lines.

Every chemical element, if it is heated in a star or a galaxy or over a Bunsen burner, will emit a pattern of distinct bright lines in a certain part of the spectrum. These patterns are very recognizable by the distance apart and the brightness of the individual lines. So when an astronomer sees that the pattern of a particular element – helium, for example, which is usually in the yellow part of the visible spectrum – has shifted its position towards the red end of the spectrum, he or she knows that the source of the helium is moving away from us and can calculate how fast it is moving by measuring how far it has shifted towards the red.

This is fine for fast-moving astronomical objects, but the use of red-shift measurements has not been so successful for objects that move or change speed slowly. There are various areas in astronomy where small changes in speed are important. One of them relates to the possibility of planets orbiting around distant stars. Such planets are not visible through telescopes but can be detected by the effect they have on the movement of the star they orbit.

A planet does not rotate around the centre of a stationary star, but instead the star and the planet rotate around a point between the centres of the two objects but nearer the more massive object, the star. It's like a drum majorette's baton with different-sized bobbles on either end, twirled round a point near one end. This means that while the planet describes a wide circle the star itself also moves in a way that gives it a blue shift and then a red shift as it approaches and recedes from the observer.

But the speed of the movements of such a star/planet system, compared with the large red shifts of receding galaxies, is so small that only the largest planets have been detected by this method, planets 300 times as massive as the Earth, where it is difficult to imagine life forming because of very strong gravitational effects.

When astronomers use a standard spectroscope to look for shifts in the spectral lines, the change in position is so small that the spectroscopes are not accurate enough to detect any movement at all. But recently, a group of scientists at the Max Planck institute in Germany invented a way of superimposing very fine calibration lines, like the marks on a metal ruler, on to the spectrum of distant astronomical objects so that they can detect even the tiniest shift, created by speeds as small as 1 centimetre per second (36 kph or 22 mph).

This new device is called a laser frequency comb, and depends on a laser that emits spectra controlled by an atomic clock, which measures time to an accuracy of 1/1,000,000,000th of a second, and can therefore produce a very precise artificial spectrum. This spectrum is then used like the markings on a ruler to fix the position of an emission line from a distant object with much greater accuracy.

The history of astronomy has advanced in major steps as new observing and measuring devices have been invented. So far, the laser frequency comb has barely been used at all to address a number of problems in astronomy that await answers, because it is still being developed. But once it is used to look for Earth-sized planets orbiting stars, it will be a major leap forward in the search for life-bearing planets outside the solar system.

LOOKING FOR A BLACK SWAN IN SPACE

The power of science to *prove* things is often overestimated. With every day that passes, and every sunrise, the hypothesis that the Earth rotates is not proved, it is merely confirmed. Alternative explanations – such as that the sun rotates around the Earth, for example – might also be confirmed by the same observation. But observations in science can be used in a more powerful way, to disprove hypotheses.

The hypothesis that 'All swans are white' is confirmed, weakly, every time we see a white swan. But one single observation – of a black (or red or blue) swan – will disprove that hypothesis.

One dramatic example of how this can be very useful in science was contained in a paper by Martin Rees and Piet Hut, two cosmologists, some years ago. They raised the possibility that the universe might be in a dangerously unstable state as a result of the cooling process that has occurred over its 13-billion-year life, [🐕 8]. Such a state is called a 'metastable minimum', because it has the *appearance* of being stable but actually might not be. It has the potential for becoming very unstable.

One way to understand this state is to imagine two mountains with a valley in between. A spherical boulder at the bottom of the valley would be stable. You could push it uphill towards either side of the valley and it would just roll back to its lowest position, however hard and long you pushed. But if one of the mountains had a shelf halfway up, with a surface that sloped down a little towards the face of the mountain, you could imagine that a boulder on that shelf would *seem* pretty stable too. If you push it towards the edge of the shelf and then stop, it will roll back to its original position. But if you push harder it will reach the edge and plummet down to the valley below. It was *metastable* and has become unstable.

For Rees and Hut, there was no way of knowing whether the universe had arrived at a stable or a metastable state after billions of years of cooling. Why would this matter?

If the universe was not truly stable but more like the boulder on the ledge, it would be possible for it to be 'pushed' off the ledge if a sufficiently large amount of energy was concentrated at one point in the universe. If that were to occur, it would trigger a wave of annihilation expanding at the speed of light which would destroy the entire universe. But how could such a concentration occur? Physicists study atomic structure by colliding atoms together at high speed in a particle collider, a construction covering several square kilometres. The collisions create new types of particles as a result [🐕 184]. To achieve this, the collision must create a concentration of very high energy at one point in space.

Rees and Hut knew on the basis of current calculations that no current particle collider could get anywhere near the danger-ous energy concentration. But the higher the energy of these devices, the more information they can gather, particularly about conditions in the very earliest moments of the universe. One day, they thought, scientists might well be able to build a collider with such high energy that it would endanger the universe.

They then reasoned as follows: if we can show that such extremely high concentrations of energy have existed some-where else in the universe at some time in the past, then, since the universe is still here, it has clearly not suffered a wave of annihilation and therefore is not in a metastable state. Effectively, they tried to find a black swan – a concentration of energy somewhere in the universe that had occurred in the past without ill effects, disproving, once and for all, the theory that the universe was at a metastable minimum.

Their calculations showed that annihilation would be triggered at a concentration above 1,000 trillion electron volts, and they looked at the universe to see whether such high concentrations existed anywhere. None of the 'normal' energy sources in the universe reached that figure. Black holes, neutron stars, white dwarfs, pulsars – all were very energetic, but not in such a concentrated way. Then they thought about cosmic rays – highly charged particles that travel at high speed through space and sometimes collide with stars or planets. But such collisions were similarly not energetic enough to cause annihilation.

Finally, they looked at one extremely unusual situation – what if two cosmic ray particles, both massive and both travelling near the speed of light, were to collide? They found that in that case, the desired – or undesired – concentration of energy could occur and if the universe were metastable it would be annihilated. The final piece of the jigsaw puzzle was to determine the chances of such a very rare event having occurred in the lifetime of the universe. Although cosmic rays collide with slow-moving particles all the time, as they enter the Earth's atmosphere, for example, the chances of one cosmic ray colliding with another are very small. Nevertheless, since the creation of the universe there has been a lot of time for such collisions to occur, and Rees and Hut worked out that there must have been a very high energy collision of two cosmic rays somewhere in the universe maybe 100,000 times in the last 13–15 billion years or so since the beginning of the universe, i.e. once every 130,000 years. The conclusion was firm and reassuring. The universe is actually stable, and there is therefore no danger of *any* concentration of energy, from particle colliders or anything else, triggering instant annihilation now or in the future.

DOES INTELLIGENT LIFE CAUSE THE UNIVERSE?

This may seem an odd question to ask, but behind it lies a complicated chain of reasoning among some scientists, one consequence of which is that the universe may only exist at all once it has conscious observers. Before that point, some presume, it was merely a mathematical abstraction.

The discussion centres on something called the Anthropic Principle. It was anticipated more than a hundred years ago by Alfred Russel Wallace, who wrote: 'Such a vast and complex universe as that which we know exists around us, may have been absolutely required . . . in order to produce a world that should be precisely adapted in every detail for the orderly development of life culminating in man.'

More recently, scientists have come to learn how improbable the chain of events is that led to the origin of life and eventually of humans, the only organisms capable of reasoning about their own origins. They point to the fact that certain physical characteristics of the universe all have the values they should have if intelligent life is to arise in the universe – its age, its expansion rate, the number of electrons and protons, even something as seemingly inevitable as the number of dimensions, three of space and one of time. But it's a slippery kind of improbability. Sometimes it can seem like the statement: 'How likely is it that at exactly 11.03 today, the phone will ring and a double-glazing salesman with the initials A.N. will try to sell me £3,500 worth of double glazing?' In fact, if you ask that question before the event, with no foreknowledge, the answer is that it is extremely unlikely. But to ask it afterwards is trivial – the answer is that the event happened and therefore it was certain.

Clearly, we are asking the question about us and the universe in a context where we know we exist – as Descartes pointed out – and therefore it doesn't seem to make much sense to raise questions about the probability or improbability of the steps that led to us being here. If the numbers had been slightly different there would have been a different universe, without us, and there would have been no one to worry about it. One hypothesis that avoids the need for this very special chain of events to have occurred in the only universe we know is that there exist parallel universes, each with a different set of laws, dimensions and constants. Those in which no form of intelligent life *could* arise, the vast majority, will clearly not have inhabitants who wonder about their existence, and a small number equal to or greater than one will have intelligent life, us. But even if ours is the only universe, the fact that we are here requires no more of an explanation than that our legs are just long enough for our feet to reach the ground or that our skeletons fit snugly inside our skin and at no point do the bones stick out.

But for the Anthropicists, if I can call them that, multiple universes are not very popular. Apart from anything else, they remove any need for their favoured hypothesis, intelligent design, and these people seem to want there to be a reason why the only universe we know has produced intelligent life on Earth. (And some of them believe this has happened only on Earth and nowhere else in the universe.) They feel there is a purpose behind the universe being the way it is. Although they do not necessarily speak of a God, in their various formulations of the Anthropic Principle they get nearer and nearer to that possibility. There is a Weak Anthropic Principle (WAP), which says the various physical characteristics of the universe had to be the way they are in order to allow carbon-based lifeforms (the only type we know) to arise and for the universe to last long enough for this to occur.

There is also a Strong Anthropic Principle (SAP), which says that the purpose of the universe is to give rise to intelligent life. This has dangerous hints of the intelligent design argument pushed by creationists in the United States.

The form of the principle that gives rise to my title is the Participatory Anthropic Principle (PAP), put forward by the physicist John Wheeler, who believes that no universe can exist unless it contains conscious observers. By observing the universe we bring it into existence.

Finally, there is FAP, the Final Anthropic Principle, following on from PAP, which says that now that life has begun it will be impossible to destroy, otherwise the universe would lose all its observers and disappear.

To give a sense of perspective, in the face of this seriously presented but largely unprovable chain of reasoning, the American writer Martin Gardner has suggested changing the name of the Final Anthropic Principle (FAP) to CRAP – the Completely Ridiculous Anthropic Principle.

WHY IS THE SKY DARK AT NIGHT?

This is one of those simple questions that scientists have asked in the past which lead to surprising and quite deep answers. Having been brought up in a world where the natural state of the night sky is darkness (unless we live north of the Arctic Circle), most people would respond to the question by asking: Why *shouldn't* the night sky be dark? Without an obvious source of brightness – the sun – it is hardly surprising that after sunset we look out into a black void punctuated only by the pinpricks of stars, and of course the moon from time to time. (We'll set aside for another time the fact that from the vantage point of the moon, even when the sun is in the sky, the rest of the sky is black.)

When this question was first asked, the universe was believed to be infinitely large, with an infinite number of stars. If that were the case, wherever you looked in the night sky your line of sight – a straight line drawn outward from your eye into space – would eventually interact with the surface of a star. It would be like looking around you in a forest with an infinite number of randomly scattered trees. Wherever you looked your view would eventually be blocked by a tree trunk.

Now if wherever you looked in the sky your line of sight met a star, you would expect the entire night sky to be as bright as the surface of a star. You might counteract that statement with the observation that 'surely the light from stars is dimmer the farther away they are?' In fact, the *total* light from a star does indeed decrease with distance, but that is because its surface is apparently smaller. Each point on its surface seen from the Earth is undimmed, but the reduced apparent size means that there are far fewer 'points' and so the total brightness is less. In an infinite universe, however, wherever you looked you would see a point of starlight, and the sky should be dazzlingly bright at night.

In the nineteenth century, at a time when scientific knowledge was not confined to scientists and nerds, a poet and novelist like Edgar Allan Poe was interested enough in the night-sky paradox to reason out an explanation and include it in what he called a prose poem:

> Were the succession of stars endless, then the background of the sky would present us a uniform luminosity, like that displayed by the Galaxy – since there could be absolutely no point, in all that background, at which would not exist a star. The only mode, therefore, in which, under such a state of affairs, we could comprehend the voids which our telescopes find in innumerable directions, would be by supposing the distance of the invisible background so immense that no ray from it has yet been able to reach us at all.

What is wrong with this reasoning? Well, nothing actually. In an infinitely large universe with an infinitely large number of randomly distributed stars the night sky would be dazzlingly bright. So we need to look and see which assumptions are wrong. You might say that perhaps light from some of the distant stars is dimmed by dust in the way. But in fact, the dust would not make the starlight disappear. There *is* interstellar dust but it is actually heated up by starlight and reradiates it, so the total amount of light is still the same. For the people who first raised this issue – Halley of comet fame was one, but the credit is usually given to Hans Wilhelm Olbers – the universe was both infinitely large and infinitely old. Any restriction of its time and space would have seemed too limiting for the universe God created. Perhaps we have to sacrifice one or both of those assumptions – infinite size or infinite age – in order to explain the darkness of the night sky.

We now believe that the universe began with a Big Bang

about 13–15 billion years ago [🐕 7]. Starting from a single point the universe expanded, very rapidly at first and then more slowly, and today it is still expanding with a 'frontier' that is about 13–15 billion light years away. Perhaps the darkness of the night sky is because, beyond that frontier, there are no stars to contribute to the brightness of the night sky, just as if the trees in the forest, that I mentioned earlier, stopped 15 miles away and we could see chinks of light beyond the edge of the forest.

As an explanation of the darkness of the sky, it's a nice try, but a British mathematician has shown that even in the limited universe we inhabit, there are enough stars scattered over the sky out to the current frontier to produce a night sky that would be blazingly bright. So we need another explanation.

The fact that the universe is expanding was not accepted until early in the twentieth century. One implication of this is that stars moving away from us seem less bright because of the Doppler effect [🐕 168]. As with the change in the frequency of sound from a moving source, the colour of light changes depending on the speed of travel, and light from a receding star becomes redder. Since the mammalian eye has evolved most sensitivity to the spectrum of colours that make up white light, as the starlight shifts towards the red, some of those colours drop out and the light seems less bright. So one reason the night sky is not as bright as we would expect is probably that the fast-receding stars that should be as bright as the slower, nearer ones are much dimmer as a result of the Doppler shift.

This explanation might be sufficient, although new know-ledge about the Big Bang and the expansion of the universe has introduced another factor. The Big Bang must have been very bright, because of the energy we know was concentrated in one spot. Shouldn't we therefore still be aware of that brightness in the night sky? As it happens, it *is* possible to detect the bright 'echo' of the Big Bang [🐕 7], but as with starlight, the Doppler effect due to the rapid expansion of the universe has shifted the colour of that early flash of light far beyond the red end of the spectrum and it's now detectable only as microwave radiation.

A deceptively simple question posed by Heinrich Wilhelm Olbers in 1823 has led scientists who have sought an answer over the last two centuries to explore a succession of new discoveries that Olbers would never have dreamed of, in their search for an explanation.

HOW LONG DOES A LIGHT YEAR LAST?

For many people, this may sound a perfectly sensible question. The expression 'light year' sounds just like a measure of time. A brief trawl through the Internet came up with the following:

> 'It seems light years ago that Scott Fitzgerald rhapsodized the lissome girls in grown-up gowns.' *People* magazine
>
> '. . .has been taken for granted in the United States for light years.' *Christian Science Monitor*
>
> 'Seville feels like light years ago; as he recalls the spectre of Spain, O'Neill knows it's time for major changes . . .' *Daily Mail*
>
> '101 light years ago' Title of rock album

'. . .the light we see from stars is out of date in the sense that it left the object often many light years ago.' Letter in *The Times*

And, although I am embarrassed to admit it, I was once responsible for a children's television programme in which the presenter said to a group of children, 'See you in a couple of light years.'

A light year is in fact a measure of *distance*, not time, so the only possible answer to my question can be: 365 days, 6 hours, 9 minutes and 9.7676 seconds – in other words, the same length as any other year.

The reason this confusing term is used is because distances to interesting objects in the universe are so huge that any multiple of an Earthly measure would be clumsy. The largest measure we use on Earth – or at least on land – nowadays is the mile, and since the nearest star – other than our sun – is about 24,689,700,000,000 miles (39,734,220,000,000 kilometres) away, it's more conveniently expressed as 4.3 light years, because it takes light 4.3 years to travel from that star. Of course, we could say 25 million million miles, which is not so cumbersome, but then when we want to talk about more distant objects, such as a far-off galaxy called IOK-1, its distance in miles is 75,715 billion billion miles, more of a mouthful than 12.88 billion light years.

If we only had tiny measurement units on Earth, if we were perhaps flu viruses measured with a unit called a micrometre, a millionth of a metre, then we'd have trouble discussing the distance from London to New York. (Of course, there might be other obstacles to such discussions – the absence of a vocal tract, for example.) If our largest unit was a micrometre, instead of talking of the distance from London to New York as

5,585,000,000,000 micrometres, we could instead talk about it as a fifth of a light second.

If you do occasionally make the mistake of thinking of a light year as a unit of time, you're in good company. Even astronomers or people who are really interested in astronomy make the same mistake. Here's another selection from the Internet, this time from sites answering astronomical queries or giving astronomical information:

'How crazy is it, though, that we can look so far into the past. . . if only we could communicate with someone there we could ask 'em to tell us what was going on here 7 billion light years ago.' Blog

'Let's say you point Hubble [telescope] in one direction and observe light from Galaxy A that departed 13.7 billion light years ago near the instant of the Big Bang, it would take light 46.5 billion light years to reach you if that light were to leave Galaxy A right now.' Astronomy buff website

'So twelve billion light years ago, that mass that we are seeing now was moving faster because it was at the brink of the big bang, if in fact the big bang happened around then.' Physics forum

'Our nearest neighbor, Andromeda, is 2.5 million light years away. Again we are not seeing Andromeda as it is now, but how it was 2.5 million light years ago.' *Scienceray*

'The light would have left Proxima Centauri 4.3 light years ago since light travels at the speed of light and the star is 4.3 light years away.' *Skywatcher*

'If our galaxy is rotating, could one see Earth millions of light years ago when it was on the opposite side of the rotation from where we are today?' Question to 'Ask an astronomer'

Although the last question was asked by an amateur, it was answered by an astronomer who didn't even point out the error in the question.

THE WORLD'S OLDEST NUCLEAR REACTOR

The average nuclear power station costs about £1.5 billion to build. At the heart of the station is the nuclear reactor, a complex and delicately engineered device with control systems to monitor the split-second events and ensure a steady generation of heat to run turbines that generate electricity. In some types of reactor, the starting material for the process that generates electricity is uranium. As everybody knows – because it is seen by some as a drawback to the use of nuclear power – nuclear power stations produce nuclear waste. This waste contains unused uranium with a very specific 'signature'. On the whole, if scientists come across this particular material they can safely assume that somewhere near by is a man-made nuclear reactor.

Except, it appears, in the central African country of Gabon. Here in 1972, geologists working in a uranium mine in a place called Oklo came across samples of uranium that had the very distinctive characteristics of nuclear waste. But there was no nuclear power station on the entire continent of Africa at the time. Something very strange must have happened.

Rock samples of naturally occurring uranium contain two types of uranium atoms, known as U^{238} and U^{235}. Most of the atoms are U^{238} but about 0.7 per cent are U^{235}. In a nuclear reactor, atoms of U^{235} are bombarded with nuclear particles called neutrons. A single neutron causes an atom of U^{235} to give off two or three more neutrons, which go on to bombard other atoms, which in turn give off more neutrons. If the conditions are right, a chain reaction is set up with an ever-increasing number of atoms disintegrating and generating heat for power generation. Among those necessary conditions is the need for what is called a moderator, usually a form of water, which acts like a blanket to make sure that too many neutrons don't escape and bring the chain reaction to a halt.

The waste from a nuclear reactor shows a much smaller proportion of U^{235} than the original 0.7 per cent, because the atoms have been broken up in the chain reaction. The rocks from the Oklo mine showed the same small proportion, as if at some point in the past a nuclear chain reaction had taken place. While most scientists were reluctant to believe this had happened, it turned out that in 1956 a Japanese chemist, Paul Kuroda, had suggested that it might have been possible in the past for a uranium chain reaction to occur naturally if certain conditions prevailed. These included a higher proportion of U^{235} than in today's naturally occurring uranium rocks, and the presence of water as a moderator to slow down the neutrons.

Scientists who studied the Oklo rocks have now come to believe that these conditions applied nearly two billion years ago, when the proportion of U^{235} was much higher, about 3 per cent. The reason today's levels are lower is that, like all radioactive elements, uranium decays into other atoms [🐕 189]. The rate at which the uranium decays is measured by a quality called half-life – the amount of time it takes a given amount of an element to reduce to half its former mass. The half life of U^{235} is 704 million years, and so several half-lives ago, round about two billion years, there would have been more U^{235} in natural deposits, just the sort of amount that would lead to a sustainable chain reaction. Provided, that is, that there was water around – the moderator – to stop the stray neutrons escaping. Which is exactly the conditions operating in Oklo at the time. What's more, the scientists who have now studied Oklo in some detail can work out that there was an interesting pattern of nuclear activity that went in cycles for millions of years. The chain reaction would start in rocks surrounded by water, the atoms would split and generate heat, the heat would turn the water to steam and destroy its ability to moderate the reactions, and the neutrons would escape, stopping the chain reaction. The steam would condense and turn back into water, beginning to blanket the neutrons that were still being emitted by the uranium, more would be retained in the sample, splitting the uranium atoms and restarting the chain reaction.

In 2004 a group of scientists in the United States worked out from a fragment of Oklo rock a few millimetres wide that the natural reactor would produce heat for about half an hour and then shut down for two and a half hours before starting up again. It had done this over a period of 150 million years at an average power of 100 kilowatts, the kind of power produced in a typical car engine.

One final pleasant surprise came out of this research – pleasant for those who worry about the future storage of nuclear waste from power stations. The nuclear waste products from the natural nuclear reactions were held in place by the surrounding rock, made up of granite, sandstone and clays. In the space of 2 billion years, this nuclear waste, including the most toxic element, plutonium, had migrated through the rock over a distance of only three metres. After the analysis of Oklo, scientists now feel more confident that the waste from modern power stations can be just as well controlled in the underground rock repositories that are planned to deal with the problem.

KILLER LAKES

During an August night in 1986, 1,700 people in villages near Lake Nyos in a mountainous region of the African state of Cameroon died in their sleep. The lake was the cause of their death, even though some people lived up to 25 kilometres from the water's edge. But that wasn't far enough away to avoid a huge cloud of carbon dioxide which bubbled up out of the lake and rolled up and down hills and into valleys, smothering all living creatures by depriving them of oxygen.

There are three extremely unusual lakes in this part of Africa. One, Lake Monoun, had killed thirty-seven people two years beforehand, but the Lake Nyos eruption was far worse. The third, Lake Kivu, has not yet erupted, although it contains just as much carbon dioxide as the others.

The deadly property of these lakes arises from the fact that they are situated in the craters of volcanoes, and over the centuries carbon dioxide seeps from the volcanic interior through the lake bed. In other, shallower lakes when this happens the gas escapes through the surface of the water, after being distributed by the natural movements of the water. But these Cameroon lakes are both stable and very deep, exerting pressure on the gas that keeps it bottled up like the water in a soda siphon. And indeed, soda water *is* carbon dioxide dissolved in water. Once in a while, perhaps after a storm or a landslide, the stability of the layer of carbon dioxide dissolved in water is disturbed and with frightening force it escapes in a huge bubble, rising hundreds of feet into the air and then, since carbon dioxide is heavier than air, it falls to the ground and rolls across the countryside. In the case of Lake Nyos, about a cubic kilometre of gas was released – 500 Wembley stadiums-full – and travelled at 60 kph outwards from the lake. If there had been detectors at the edge of the lake, the people farthest away who died would have had about fifteen minutes' warning and some of them might have been able to escape the disaster.

As a result of these events, which occurred without any warning, French scientists have now installed a polythene pipe which goes 200 metres down to the lake bed and allows carbon dioxide to escape before it builds up to dangerous levels. The scientists, who are based in Paris, monitor the lake by satellite, and can open valves in the pipe if, in spite of the release of gas, there is a build-up to dangerous levels.

Unfortunately, the investigation of this unusual lake has led to the discovery of another potential hazard. There is a weak dam at its northern edge, and if it ruptured it could lead to floods and a gas eruption that might drown or suffocate up to ten thousand people.

And the news continues to be bad. Although the degassing pipes have helped, a team that visited Lake Nyos in 2006 reported that the lake is still full of gas that could erupt at any time, posing 'grave dangers to local populations'.

ON THE ROCKS

One of the consequences of severe global warming will be a rise in the sea level around the world, as an increase in temperature melts the polar ice caps. As most people know, the north and south poles are both covered with ice, and there are already signs of shrinkage over the last few years suggesting that the process has begun. The Arctic ice cap has shrunk by 20 per cent in the last thirty years. But the problem is only half as bad as it might seem, at least as far as a change in the sea level is concerned. If both polar ice caps were to melt entirely, only the melted ice from the South Pole would have an effect on the sea level.

It's not that there's anything different physically about the actual ice at the two poles and the water it turns into. But there is a fundamental difference in what lies underneath the two ice caps. The North Pole is a large slab of ice floating in water; the South Pole is a large slab of ice resting on land. To understand why this makes such a difference, we can take the example of an iceberg. As water freezes, the resulting block of ice is a little less dense than the water it came from. When molecules of H_2O form a liquid – water – each molecule is bound loosely to an average of 3–4 other molecules. As the temperature drops and ice is formed, each molecule becomes bound more rigidly to four others, in such a way that the molecules are more spaced out. So ice floats in water, because a volume of ice is less dense – more spaced out – than an equal volume of water. (Just as a block of lead *sinks* in water because an equal volume of lead is more dense than an equal volume of water.)

So what happens when a block of ice floating in the sea melts? Because ice is less dense than water – i.e. occupies less space for its weight – the volume of the water that comes from the melting of the *whole* iceberg is equivalent to the volume of the portion of the iceberg beneath the surface when it was still ice. So the actual volume of the sea, which determines the sea level, is unchanged. All the icebergs in the sea and all of the ice at the North Pole could melt overnight and it wouldn't affect the sea level at all. (In the same way, if you put ice cubes in half a glass of water and then fill the glass to the top, as the ice melts the glass will not overflow.)

In the case of the South Pole, however, the ice is not in the sea at all. It is firmly on solid land, the continent of Antarctica. Once that starts to melt, the water produced will pour into the southern seas that surround the land, and begin to raise the sea level. If the whole Antarctic ice sheet were to melt, it could raise sea levels by at least thirty feet (ten metres) and flood many low-lying parts of the world.

None of this is meant to suggest that we don't need to worry about the melting of the North Polar ice cap, which is happening more rapidly than many scientists expected. There are other effects on the delicate balance of the global ecology. The contraction of the Arctic ice cap, even though it doesn't raise the sea level, will reduce the amount of sunlight that is reflected away from the Earth, and this light and heat will be absorbed instead, raising the temperature of the Earth.

It all makes you want to add another ice cube to your drink and take your mind off things by watching a disaster movie.

UP FROM THE ASHES

The ashes of a beloved parent or other relative seem harmless when taken to the relative's favourite mountain or beauty spot and dispersed with a small ceremony of remembrance. But with 70 per cent of people being cremated, and most of us wanting to respect the deceased's wishes for a calmer resting place than a corner of a family living room, the beauty spots of the UK are suffering from the effects of a chemical imbalance caused entirely by the massed ashes resulting from cremations.

The situation has become so serious on the peak of Britain's highest mountain, Ben Nevis, that the authorities have asked people not to scatter ashes there. The high levels of calcium and phosphorus in the ashes are changing the chemical balance of the soil and affecting the alpine plants that tolerate only a narrow range of soil acidity or alkalinity and already struggle to survive.

Ironically, while some plants are killed by the ashes, others flourish and expand into areas that were previously barren. The effect is similar to gardeners putting coal ash or bone-meal on vegetables. At higher altitudes the ash promotes the growth of mosses and grasses which cover rock and soil that were formerly bare.

One scientist commented that the ashes were a 'bean feast' for some plants, and claimed that on one Highland mountain he had been monitoring, the scattered ashes from one crem-ation sixteen years ago had produced visible and long-lasting changes in the plant life.

MR GLOBAL CATASTROPHE

The threats to our environment – global warming, environmen-tal pollution, and so on – seem so huge and widespread that the idea that any one individual has made a significant contribution to the parlous state of the world's environment seems unlikely. Nevertheless, one man, an American inventor, was responsible for two of the most insidious contributions to the global en-vironment – lead in petrol and the refrigerant freon.

Thomas Midgely Jr suggested the use of lead as an additive to petrol in order to prevent a problem known as 'knocking',

whereby the fuel would burn too early in the internal combus-
tion cycle and lead to noise and engine damage. Even at the
time of the discovery, lead was known to be harmful to
humans and the manufacturers called it 'ethyl' instead of
using its full name, 'tetra-ethyl lead'. After a few years
working with the additive, Midgely himself suffered the ill
effects of his invention and developed lead poisoning. This
was only a minor setback. In spite of his own experience and
the death of ten workers at the manufacturing plant, along
with hallucinations and insanity in others, Midgely tried to
reassure people by pouring the additive over his hands and
breathing it in. He claimed that he could do this every day
without harming himself, but there is no evidence that it
became part of his daily routine.

His next service to mankind was to devise a compound of
various atoms – chlorine, fluorine and carbon – as a replace-
ment for substances used in refrigerators at the time which
were highly toxic to humans if they leaked out. Freon, also
called dichlorodifluoromethane, was seen as a miracle compound. It was colourless, odourless, non-flammable, non-
corrosive and seemed generally a harmless fluid that would sit
quietly in your refrigerator changing from liquid to gas and
back again as it circulated through the system and kept your
food cool.

Within a few years of being introduced, freon was the standard coolant for domestic refrigerators. As seemed customary
with his more noxious inventions, Midgely conducted a public
demonstration of freon's low toxicity and generally benign
nature by inhaling a lungful of the gas and breathing it out on
to a candle flame, which was thereby extinguished.

Both of Midgely's inventions made fortunes for the companies that manufactured and marketed them, and dominated

the field for decades. But first tetra-ethyl lead and then freon and similar refrigerants were found to have had disastrous effects on the environment. Lead in the atmosphere released by car exhausts entered the blood and caused neurological damage in children; the chemicals released from discarded refrigerators made a major contribution to the hole in the Earth's ozone layer.

But Midgely was long dead by the time the harm caused by his inventions was realized. There was a kind of poetic justice in the way he died – as a result of yet another of his inventions. Midgely contracted polio in 1940 and lost the use of his legs. He invented a special harness to get himself out of bed but his ingenuity failed to foresee a hazard of the apparatus, and when trying to get up on 2 November 1944 he was strangled by his own invention.

PLUS AND MINUS

THE MOST IMPORTANT MACHINE THAT DOESN'T EXIST

Alan Turing was a British mathematician whose creative thinking laid the groundwork for modern digital computers. His fame rests partly on the 'Turing machine', which never existed outside his brain or the scientific papers he wrote. But the way computers work today depends on the insights of Alan Turing and a small group of like-minded thinkers in the 1930s.

Turing was trying to answer a question posed in 1928 by the German mathematician David Hilbert. Hilbert asked whether there was a mathematical procedure that could determine in any mathematical system whether a statement in that system was true or false. Turing ended up showing that there were some systems, arithmetic is one, where it is impossible to decide using only techniques within that system whether its statements are true or false.

In a scientific paper exploring this problem, Turing devised an imaginary machine, as part of what scientists call a 'thought experiment'. His machine consisted of an endless tape divided into cells in which symbols could be written or erased by a head, like a magnetic-tape head.

In Turing's paper he imagines the changes in the cells being carried out by what he called a 'computer' (in those days the word referred to a human being): 'Computing is normally done by writing certain symbols on paper. We may suppose this paper is divided into squares like a child's arithmetic book . . .

The behaviour of the computer at any moment is determined by the symbols which he is observing and his "state of mind" at that moment.'

The simplest repertoire of symbols would be 0 and 1, and accompanying this 'machine' would be a table of instructions. Such a table might include instructions like the following:

> If the cell under the head contains 0, erase it and print 1 and then move the tape to the right.
>
> If the cell under the head contains 1, erase it and print 1 (again) and then move the tape to the left.
>
> If the cell under the head contains 0, erase it and print 1 and then move the tape to the left.
>
> If the cell under the head contains 1, erase it and print 1 (again) and then move the tape to the right.
>
> If the cell under the head contains 1, erase it and print 1 (again) and then leave the tape where it is.

These instructions (just part of a complete table) could be summarized briefly as:

$$(0,1,R), (1,1,L), (0,1,L), (1,1,R) \text{ and } (1,1,N)$$

This table of instructions applied over and over again will lead from the initial state of the 'machine' – one arrangement of symbols – to some final state. With the right set of instructions, the starting state of the tape – say, a representation of the number 27 – could lead to a finishing state, 729, using a table of instructions that multiplies numbers by themselves.

Having invented in his head a Turing machine that would solve a single problem using a table specific to that problem,

Turing then showed that it would be possible to invent a Universal Turing machine that could simulate all other Turing machines. The 'table' for that machine was equivalent to the software in modern digital computers that allows them to be used in a huge variety of ways.

Although this Turing machine was not built, he was no slouch when it came to building real machines for solving problems. One important mathematical problem Turing attempted to solve, still unsolved today, was a proof of a mathematical statement called the Riemann hypothesis, which dealt with the distribution of prime numbers.

In 1939 he received a grant to construct a machine that consisted of thirty meshing gear wheels with different numbers of teeth related to logarithms. Each wheel had a weight attached at a certain distance from the centre, and the wheels were arranged in meshing groups set in motion by the turning of a large handle.

Turing's biographer, Alan Hodges, wrote:

> [Turing's] room, in the summer of 1939, was liable to be found with a sort of jigsaw puzzle of gear wheels across the floor . . . Alan tried and lamentably failed to explain what it was all for. It was certainly far from obvious that the motion of these wheels would say anything about the regularity with which the prime numbers thinned out, in their billions of billions out to infinity.

Although his Riemann machine never worked, Turing did help to develop one of the most epoch-changing machines ever built, for decoding the German Enigma code during the Second World War. This work was said to have shortened the war by two years and it earned him an OBE.

Pi = 3

Everyone has heard of the number pi, represented by the Greek letter π, but most of us are unfamiliar with its peculiar nature.

Its origins are not particularly mysterious. Early mathematicians, including Egyptian, Indian, Babylonian and Greek geometers, discovered that every circle had the same ratio between its circumference and its diameter. Whether it's the size of a penny or a circle with the diameter of the orbit of Pluto, that ratio was exactly the same, a number approximately equal to $3^1/_7$. So the circumference of a circle is always a little over three times the diameter.

Knowledge of this ratio was useful if you wanted to draw a circle on the ground with a specific circumference, say 10 metres, using a peg, a rope and a piece of chalk. The length of the rope had to be a little under a sixth of the circumference of the circle, 1.6 metres in this case, because the radius is half the diameter.

As circles were measured with greater and greater degrees of accuracy, the estimate for this ratio, π, became more and more precise. The Egyptians used 25/8, the Babylonians 256/81, and by 2009, using computer methods rather than tape measures round very large circles, π had been calculated to 1,240,000,000,000 decimal places, a seemingly random sequence of the digits from 0 to 9. That is, a figure that starts 3.1416 and goes on for another 1,240,000,000,000 figures. And like Jorge Borges' Library of Babel, these figures, if extended to infinity, contain any combination of digits you care to seek. My birthday, for example, in numerical form, starts at position 36,764,575, and my surname, where A = 1, B = 2 and so on, starts at position 82,062,313. (This last number, by the way, turns out to be prime, a matter of some pride to my family.)

Now here's the peculiarity. Ordinary numbers aren't meant to go on like this. If I measure my height with very great accuracy it comes to 180.236128639 centimetres. And then it stops. Or rather, if I try to add more digits they will all be 0. (This example is slightly tongue-in-cheek.) If we turn the Egyptian 25/8 into a decimal it becomes 3.125 and then stops. You can write it as 3.125000000000 with a trillion more zeros but it won't do you much good and you'll get writer's cramp. There won't be non-zero digits lurking in the billionth position.

π is a member of the class of *irrational* numbers, given that name not because they behave in an irrational way but merely because they cannot be expressed as a *ratio* of two whole numbers. It's actually a member of a smaller group within the irrational numbers, called transcendental numbers, and although there are not very many members of this group known, it turns out that they far outnumber all the other numbers we are more familiar with – whole numbers, fractions and so on.

For people of a non-mathematical bent, all this complexity can be worrying, particularly when, like π, numbers relate to commonplace parts of our real world like circles. The idea of a number that can never be written down exactly and yet is all around us in the coins we use, the sun we look at, the steering wheel we hold seems unnatural.

Which is why in Indiana in 1897, a member of the state congress decided to deal with the problem by legislating the value of π to something more reasonable, more rational, you might say. T. I. Record, of Posey County, proposed a bill with a choice of values of π, all of them much more useful than the irrational one. The bill declared, 'Since the rule in present use fails to work . . . it should be discarded as wholly wanting and misleading in the practical applications.' The inhabitants of Indiana were given a choice of values. The two simplest were 4 and 3.2, but rather oddly in view of the drive towards simplification the third option was the square root of $2 \times 16/7$, about 3.23.

Bearing in mind the predominance of fundamentalist Christianity in the American Midwest, perhaps Representative Record had recourse to the Old Testament to support his bill. In the first book of Kings it is written: 'And he made a molten sea, ten cubits from the one brim to the other: it was round all about, and his height was five cubits: and a line of thirty cubits did compass it about.'

In other words, God herself said that π is exactly 3.

STUFFED IN HASTINGS

For thirty years or so around the turn of the nineteenth century, a Hastings gun dealer and taxidermist, George Bristow, was the source of a series of reports of rare bird sightings over parts of Sussex. In the way that things were done in those days on the basis of Bristow's word alone, the *Handbook of British Birds* included details of these sightings in its annual list of rare birds seen in Britain. It was enough that the taxidermist had submitted the skins or stuffed corpses of dead birds, along with some indication of where they had been seen and shot.

The first to be recorded was a woodchat shrike in 1892, and the last a masked wagtail in 1919. In the most fertile period of rare bird-spotting over Hastings, forty-nine species were added to the master list of birds seen in the British Isles, of which thirty-two were reported from the Hastings area alone.

It was an astonishing haul, and there were three possible explanations. Either Hastings and its surrounding area were a kind of reverse 'Bermuda Triangle' for birds, where rare species appeared at a far greater frequency than elsewhere in the British Isles; or Bristow was an extremely skilful and assiduous birdwatcher; or he was reporting birds that had never been seen alive in the British Isles but had somehow been acquired by him from much farther afield.

In 1962, these birds, known by then as the 'Hastings Rarities', were still included in the authoritative list of birds known to have visited the British Isles, and it was decided to substantiate once and for all the belief that had been growing in many people's minds that the Bristow reports were actually fraudulent. One pair of ornithologists wrote in *British Birds* in 1962, perhaps over-diplomatically, that in considering one of the Hastings 'sightings', 'we find ourselves forced uncomfortably close to incredulity'. They also commented on the 'chronic lack of fortune among expert ornithologists in seeing alive any significant fraction of the exciting rarities whose corpses were being so efficiently and profusely collected and brought in by gunners from a couple of dozen or so parishes'.

There was much circumstantial evidence that the Hastings Rarities were fraudulent. The reports had dwindled and finally stopped at the end of the 1920s, and by comparing later sightings of similar birds in the rest of the British Isles it was possible to show two things:

1. A comparison of the list of rare birds spotted in *later* years (i.e. not the Hastings Rarities) with the incidence of the same birds in the period of the Hastings Rarities showed a similar distribution of bird species, although

in greater numbers because of the improved techniques of bird-spotting and recording and the larger numbers of bird-spotters.

2. Comparing the list of rare birds spotted in the *earlier* period with those recorded later showed a dramatic difference, with many species recorded earlier which were never spotted again later, anywhere in the British Isles, let alone over Hastings.

Clearly, Bristow was reporting a whole range of bird types during the period when he was active that were never seen again by later birdwatchers, a pretty convincing sign that something underhand was going on.

But the clinching factor against the genuineness of the Hastings rarities was a simple mathematical test, called 'chi-squared', used in many basic statistical analyses of experimental results in science.

Here's how it works. Supposing you toss a coin 100 times and you get this number of heads and tails:

	Heads	Tails	Total
Observed	53	47	100
Expected	50	50	100

You want to know whether this distribution, a little skewed in favour of heads, is likely to have happened by chance, or if the coin is dodgy.

The chi-squared test consists of subtracting the expected result, E, from the observed result, O, squaring it (multiplying it by itself) and dividing it by E. This is done for both heads and tails and added together. In this case, chi squared equals $(53-50)^2$ divided by 50, plus $(47-50)^2$ divided by 50, i.e. $9/50 + 9/50$, which equals 0.36. Using a standard chart, you can look up the chi-squared value and see whether it comes within normal limits – that's to say, see whether it represents a result that could occur with random coin-tossing. In this case it does. The chart tells us that this result would occur about 55 per cent of the time, not at all unusual.

Supposing instead the distribution was, say, heads 40, tails 60. Then chi squared would be $100/50 + 100/50 = 4.0$. Now, looking up the chart, we find that the probability of this happening with a normal coin is about 4 per cent. This is not impossible, but certainly a reason for suspicion.

The discoveries in the Hastings area were clearly different from those in the rest of the country, by virtue of the fact that there were many more rare birds reported during the earlier part of the century. But if these were genuine sightings and there were more of them because Hastings had a particularly skilled group of birdwatchers or gunmen or both, then the *pattern* of discoveries in the Hastings area – rare birds of different types, seasons when more birds were spotted, and so on – would be similar to that found in other parts of the country, even if the numbers were greater. But if the rare birds were coming from some fraudulent source and were nothing to do with what was actually flying over the Hastings area at the time, the pattern would be different. A statistician who worked with the ornithologists on the *British Birds* report gathered data from three areas and two periods. He looked at reports of three categories of rare birds from each area, called Class 1, Class 2 and Class 3, and compiled a table showing the numbers in each category from the Hastings area, and the numbers from the other areas combined. This is what he found:

	Class 1	Class 2	Class 3	Total
Hastings	243	208	165	516
Remainder	125	119	255	499

Faced with these data, anyone can see that they *look* funny. The proportion of Class 1 rarities among the Hastings reports is twice that for the rest of the country. Conversely, as a proportion there are far *fewer* Class 3 rarities in Hastings.

But it was the simple chi-squared test which clinched it. As we see in the example of coin-tossing, a chi-squared result of 4 shows that there are only four occasions in a hundred where such a result could arrive by chance. In the case of the Hastings Rarities, the chi-squared test gave a far larger and far less probable result of 57.40, removing any possibility that these reports were the result of the usual processes of bird-spotting and reporting. Clearly, the 'uncomfortable closeness to incredulity' on the part of British ornithologists was entirely justified.

George Bristow stuck to his guns – so to speak – all his life, but it is significant that once the British ornithological community started to question his reports, and impose conditions on them, the flocks of rare birds that had thronged to the Hastings area dwindled to the same numbers that were spotted over the rest of the country.

So what was actually going on? No one ever proved it, but the theory hinted at in *British Birds* was that Bristow had a scheme whereby sailors regularly visiting a nearby port were encouraged to shoot birds in foreign parts, put them in the ship's refrigerator, and bring them to Bristow. He would give them a good price and stuff the birds, reporting one specimen to the *Handbook of British Birds,* and selling others on to rare bird collectors. Clearly what was commonplace in North

Africa, say, would only be rare if it was spotted over Hastings, and so, by reporting the birds as seen and shot in England, he could justify a higher price to the collectors.

ROPE AROUND THE EARTH

If you could stretch a rope all the way round the Earth so that it fitted snugly around the equator, how much longer would you have to make the rope in order to be able to raise it 3 feet, or 1 metre, above the surface?

Intuitively, one's first guess is that to raise the rope a metre high all around the Earth, you'd have to do some calculation involving the length of the rope – i.e. the circumference of the Earth. But if you're told that the rope when it fits snugly is about 40,000 kilometres long, how does that help? It certainly suggests that you'll have to lengthen the rope by some number of kilometres to be able to create an amount of slack all the way round. But supposing you're told that the answer has nothing to do with the length of the rope?

Finding the answer comes down to finding the difference between *two* circumferences – the circumference of a circle with the diameter of the Earth, and the circumference of a circle with the diameter of the Earth plus 2 metres (one at either end). If we call the first C_E and the second C_{E+} we need to know one other thing. The circumference of any circle is equal to the diameter of the circle multiplied by a fixed number, known as π [🐕 52], which is about 3.14. So we can say that $C_E = 3.14 \times D_E$ and $C_{E+} = 3.14 \times (D_E + 2)$, where D_E is the diameter of the Earth. If we are trying to find the *extra* length of the rope, we need to subtract C_E from C_{E+}. So we need to subtract $3.14 \times D_E$ from $3.14 \times (D_E + 2)$. The second expression, by removing the brackets, becomes $3.14 \times D_E + 3.14 \times 2$. If we write it out like this, we can see what the answer is:

The extra length of the rope $= 3.14 \times D_E + 3.14 \times 2 - 3.14 \times D_E$.

Or, by rearranging slightly, $3.14 \times D_E - 3.14 \times D_E + 3.14 \times 2$. While this is not Stephen Hawking-level maths, playing around with pluses, minuses, brackets and equals is not an everyday occupation for some of us, but it becomes clear that the rope has to be lengthened *not* by hundreds of kilometres, *not* even by a single kilometre, but by about twice 3.14 metres, or 18 feet.

Since we didn't need to use the actual length of the rope in our calculations, this means that *any* circular rope of *any* size needs to be lengthened by only 3.14 metres to make its circle a metre larger in diameter. So the same increase – 3.14 metres – will be necessary for a rope stretched around the base of the dome of St Paul's (110 metres) as for a rope stretched around the orbit of Jupiter (about 5,000 million kilometres).

MOZART'S DICE WALTZ

If we are to believe the music publisher J. Hummel of Berlin, the composer Theophilus Mozart (see below) dabbled in probability theory by composing a piece of music whose final shape was determined by throws of dice.

Hummel published a table of bars of music in 1793, two years after Mozart's death, which he said the composer had compiled with the intention of generating an extremely large number of versions of the 'Dice Waltz', with the participation of an audience. The table consisted of a set of 171 bars of music divided into sixteen sections of eleven bars. Each of the sixteen sections had eleven possibilities, and by throwing a pair of dice to produce the numbers from 2 to 12, people in the audience compiled a sequence of numbers that were then used to determine which version of each bar was played. If the throws of the dice produced the sequence 3, 8, 9, 6, 3, 4, 2, 7, 5, 8, 8, 12, 10, 4, 7, 6, after subtracting 1 from each number (because two dice will never produce the number 1), the musicians would start the waltz by playing the second variant of bar 1, the seventh variant of bar 2 and so on. In this way, no performance would ever be identical. Since there are 759,499,667,966,482 ways of arranging the numbers from 1 to 11 (or 2 to 12), it's unlikely that more than one in a million of the variants has ever been heard. To play them all would take over 500 million years.

And if you thought Mozart's Christian name was Wolfgang Amadeus, you might be surprised to learn that this wasn't always the case. When his talent was revealed to the world as a child of eight, an antiquarian and polymath called Daines Barrington wrote an account of tests on the prodigy carried out in London and illustrated it with a picture of the boy labelled 'Theophilus Mozart'. What's more, there is only one document in Mozart's lifetime that uses the name 'Amadeus', a Latinized version of Theophilus. He was actually christened Joannes Chrysostomus Wolfgangus Theophilus, with no 'Amadeus' in sight.

THE ULTIMATE 'AIDE-MEMOIRE'

In 1945, in the *Atlantic Monthly*, Vannevar Bush, an American engineer, wrote an article describing some ideas for peaceful technology which scientists should be working on, after five years of warfare research culminating in the atomic bomb, which he had helped to develop.

One 'invention' in particular had a quaint mixture of accurate prophecy and wide-of-the-mark clunkiness. Prediction is never easy, particularly of the future, as Niels Bohr said, and in a cartoon published at the end of the eighteenth century, showing life in the year 2000, one of the most adventurous suggestions for future advancement was written on the side of a flying delivery van (held up by balloons, the only form of flight known at the time) which was advertising 'cast-iron glass'. In fact, the glass in our windows still breaks and we've managed to live with that. But at least Bush's suggestion for a new technological device is finally coming to pass, if not quite in the way he envisaged it.

Bush prophesied something he called a 'memex'. 'A memex,' he wrote, 'is a device in which an individual stores all his books, records, and communications, and which is mechanized so that it may be consulted with exceeding speed and flexibility. It is an enlarged intimate supplement to his memory.'

So far it sounds a bit like a combination of an iPod, an eBook reader and some kind of digital recorder, all pretty prophetic for sixty years ago. In fact, as you read on, you realize that Bush's way of achieving this ambitious aim is very different from how things have turned out:

'It consists of a desk,' Bush continued,

> and while it can presumably be operated from a distance, it is primarily the piece of furniture at which [the user] works. On the top are slanting translucent screens, on which material can be projected for convenient reading. There is a keyboard, and sets of buttons and levers.

Otherwise it looks like an ordinary desk. In one end is the stored material. The matter of bulk is well taken care of by improved microfilm. Only a small part of the interior of the memex is devoted to storage, the rest to mechanism.

Bush described a series of tasks this 'desk' could do, including the kind of hyperlinking and searching that we now use the World Wide Web to do. But in his case it was done with levers, photographs and text reduced to microfilm, and whirring cog wheels to access many thousands of pages of microfilm that would contain the user's 'books, records and communications'.

Nowadays, much of what Bush's 'memex' would do is achievable with technology considerably smaller than a desk, and many of us have such devices. But one of Bush's suggestions is not quite here yet, although the signs are that it won't be long before it is. He wrote about storing and retrieving an individual's *communications*, and – apart from emails – most of us communicate by speech in a way that is not yet stored and retrievable, except, of course, by the secret services of our countries.

Yet the time cannot be far off when, if we wish it, everything we say to anyone at any time of the day can be recorded, stored, transcribed and analysed on a portable device. Of course, the more information we gather, the more important it is to have an effective method of retrieval. In 2002 scientists at the Carnegie Mellon University devised and tested an apparatus for recording every single conversation the wearer took part in, and then retrieving it later in a way that would be particularly useful for people who are not very good at remembering faces, or at least assigning names to them.

The equipment consists of two microphones and a miniature camera, worn in the lapel and connected to a laptop computer worn on the back. One microphone is directional and receives the wearer's voice, the other covers a wider area and can record the other speaker in the conversation. (Only one side of the conversation is recorded in full, because in the USA it is illegal to record a conversation without the permission of both parties.) The interesting part of the apparatus is the camera. This is used not so much to capture a video of the conversation as to obtain an image of the other party's face.

The visual and audio information about the speaker's face and voice is stored along with the wearer's side of the conversation on the laptop. The use of this equipment that the scientists had in mind was to help the wearer's memory the next time he met the person he'd spoken to. The computer would capture a picture of the face of the new speaker and compare it with the faces of all the people the wearer had ever spoken with previously. It would also compare a short clip of the new speaker's voice with clips of the voices of all previous conversationalists. These two methods, face and voice, the scientists say, would identify with a high degree of probability whether the new speaker had been encountered before. If he or she had, software in the computer would retrieve the earlier conversation and summarize it, providing a brief playback – presumably discreetly – so that, in the words of their report, it could 'overcome age and other limits to mental capacity and help recall the details needed in a given situation'.

Even in the few years since the Carnegie Mellon scientists published their report, miniaturization of computer memory along with ever more sophisticated software for facial and voice recognition has continued inexorably. The day is not far off when – if we want this – *all* of our daily activities could be recorded and stored in vision and sound for as long as we like.

There is, of course, a danger with this approach. As anyone knows who has lost a personal organizer or a computer, as we store more and more facts about aspects of our lives in artificial memories outside our brains we come to rely less and less on our own internal memories.

Plato wrote of a similar danger some years before the invention of the digital computer, when one of his characters expresses a worry about the invention of writing. In *Phaedrus*, the Pharaoh Thamus says of writing, 'If men learn this, it will implant forgetfulness in their souls: They will cease to exercise memory because they rely on that which is written.'

DOES 1 + 1 REALLY EQUAL 2?

Mathematicians take nothing for granted. To arrive at a complicated conclusion you have to make sure that every step of the way, starting from even the simplest of starting points, is correctly and rigorously argued. In the case of that tricky question 'Does 1 + 1 equal 2?', you have to start by understanding what 1 means, then go on to consider what 2 means, and finally establish that a union of 1 and 1 is identical to what you understand by 2.

These musings are prompted by the massive and important three-volume work *Principia Mathematica,* by Alfred North Whitehead and Bertrand Russell, published between 1910 and 1913, in which at least half a page is given to the proof that 1 + 1 = 2. (And half a page is a conservative estimate – one mathematician has written, 'Whitehead and Russell's Principia Mathematica is famous for taking a thousand pages to prove that 1 + 1 = 2.')

Russell himself, confronted by people who saw no necessity to prove basic arithmetical identities, wrote:

> You might say, 'none of this shakes my belief that 2 and 2 are 4.' You are quite right, except in marginal cases – and it is only in marginal cases that you are doubtful whether a certain animal is a dog or a certain length is less than a metre. Two must be two of something, and the proposition '2 and 2 are 4' is useless unless it can be applied. Two dogs and two dogs are certainly four dogs, but cases arise in which you are doubtful whether two of them are dogs. 'Well, at any rate there are four animals,' you may say. But there are microorganisms concerning which it is doubtful whether they are animals or plants. 'Well, then living organisms,' you say. But there are things of which it is doubtful whether they are living organisms or not. You will be driven into saying: 'Two entities and two entities are four entities.' When you have told me what you mean by 'entity', we will resume the argument.

It was an attempt to define 'entity' that took up some space in Russell and Whitehead's proof that $1 + 1 = 2$:

$*54\cdot43$. $\vdash :. \alpha, \beta \, \epsilon \, 1 . \supset : \alpha \cap \beta = \Lambda . \equiv . \alpha \cup \beta \, \epsilon \, 2$

Dem.

$\vdash . *54\cdot26 . \supset \vdash :. \alpha = \iota^{\prime}x . \beta = \iota^{\prime}y . \supset : \alpha \cup \beta \, \epsilon \, 2 . \equiv . x \neq y .$

$[*51\cdot231]$ $\qquad\qquad\qquad\qquad\qquad\qquad\qquad \equiv . \iota^{\prime}x \cap \iota^{\prime}y = \Lambda .$

$[*13\cdot12]$ $\qquad\qquad\qquad\qquad\qquad\qquad\qquad \equiv . \alpha \cap \beta = \Lambda \qquad (1)$

$\vdash . (1) . *11\cdot11\cdot35 . \supset$

$\qquad \vdash :. (\exists x, y) . \alpha = \iota^{\prime}x . \beta = \iota^{\prime}y . \supset : \alpha \cup \beta \, \epsilon \, 2 . \equiv . \alpha \cap \beta = \Lambda \qquad (2)$

$\vdash . (2) . *11\cdot54 . *52\cdot1 . \supset \vdash . \text{Prop}$

From this proposition it will follow, when arithmetical addition has been defined, that $1 + 1 = 2$.

(Even this proof only works 'when arithmetical addition has been defined', and that's a whole other issue.)

One mathematician has redefined what Russell and Whitehead were trying to prove in words rather than the symbols they use above: 'It asserts that if sets α and β each have exactly one element, then they are disjoint (that is, have no elements in common) if and only if their union has exactly two elements.'

While this makes the proof slightly less impenetrable it also suggests why it needs a bit of work to complete it. 'Set theory' was initiated in the late nineteenth century as a foundational system for mathematics, starting with the concept of a 'set' as a collection of objects and considering rules for combining sets and analysing relations between them. For example, the expression '*11.54' above refers to a statement elsewhere in the book that says that 'you can take an assertion that two things exist and separate it into two assertions, each one asserting that one of the things exists'. The ordinary numbers and the way we use them are just a pale shadow of the edifice erected by mathematical philosophers like Russell and Whitehead.

But it's easy to see why rigour is necessary. Sometimes, our 'ordinary' way of looking at things, even in school-level maths, can lead us astray. Here's a proof that $3 = 4$:

Suppose:
$a + b = c$
This can also be written as:
$4a - 3a + 4b - 3b = 4c - 3c$
(Because $4a - 3a$ is just 'a', $4b - 3b$ is just 'b' and so on.)
We can reorganize the terms in a different order.
$4a + 4b - 4c = 3a + 3b - 3c$

(You are allowed to move terms from one side of the equals sign to the other as long as you change plus to minus and vice versa. So, for example, $4x - 3 = 0$ can be rewritten as $4x = 3$, by moving across -3 and changing it to $+3$. This is equivalent to adding the same number, $+3$, to both sides of the equals sign. If you add the same amount to two equal quantities they remain equal.

Now we can rewrite this as:
4 times $(a + b - c) = 3$ times $(a + b - c)$
Divide both sides by $(a + b - c)$ and we discover that
$$4 = 3$$

This fallacy turns on a common mistake anyone who is not finely attuned to the rules of arithmetic can make. Faced with such trickery, for many of us it is tempting to trust common sense rather than the rarefied heights of mathematical reasoning. We are like Mrs La Touche, for example, a Victorian woman about whom the only thing that is known is that she once said:

I do hate sums. There is no greater mistake than to call arithmetic an exact science. There are permutations and aberrations discernible to minds entirely noble like mine; subtle variations which ordinary accountants fail to discover; hidden laws of number which it requires a mind like mine to perceive. For instance, if you add a sum from the bottom up, and then from the top down, the result is always different.

IT ALL BEGAN WITH A DINNER PARTY. . .

Mathematics has a way of getting complicated very quickly. You start with a simple problem that everyone can understand and before you know it the entire thing has got out of hand and your brain hurts.

One example of this is the following: at a dinner party with six guests, *either* three of them are already known to each other *or* three of them are complete strangers to each other. Prove it.

It sounds plausible but the answer keeps slipping away as you think about it. It's not saying that the group divides into two lots of three – friends and strangers. Nor is it saying that they can't all be friends or all be strangers. And on one level it might seem obvious – if two people are friends then the other four are strangers and vice versa. But that's not true either. Two of the 'strangers' may know each other without knowing either of the two 'friends'.

A mathematician, however, would do away with all that muddled thinking. He'd take out a pencil, or rather two pencils, red and blue, or perhaps three pencils, red, blue and black, and draw six black dots in a rough circle to represent guests at the dinner. Then he would draw a red line between every pair of people who know each other and a blue line between every pair of strangers. In that pattern of fifteen lines, it is possible to prove that there will always be either a red triangle or a blue triangle – three people who know each other or three people who don't.

Now of course the drawing doesn't prove this, but it has converted a fuzzy human problem into a precise mathematical statement. The problem now is phrased in terms of dots connected by lines, a graph, rather than people and their social relationships, and simple geometry can provide the proof.

The field that tackles such problems is called Ramsey theory, after a brilliant Cambridge mathematician, Frank Ramsey, who died at the age of twenty-seven, having made major contributions to mathematics, economics and philosophy. The dinner-party problem is one of the simplest in the field, and more complex Ramsey problems deal with graphs with more points joined by more lines. A graph in which all points are connected to all others by straight lines is called a complete graph. Any graph contained within that nest of lines, such as the red and blue triangles above, is called a subgraph. Ramsey problems ask questions like: what is the minimum number of points such that a complete graph randomly coloured with red or blue pencils will contain either a red triangle or a blue quadrilateral?

These problems are surprisingly difficult to solve. The Ramsey theory equivalent of the dinner-party problem rephrased for five friends or five strangers has never been solved. The answer is defined as $R(5,5)$ – the smallest number of people necessary to ensure that either five are friends or five are strangers – but no one knows what the answer is. The nearest mathematicians have come to the answer is to say that it is between 43 and 49. The Hungarian mathematician Paul Erdös once wrote:

> Imagine an alien force, vastly more powerful than us landing on Earth and demanding the value of $R(5,5)$ or they will destroy our planet. In that case, we should marshal all our computers and all our mathematicians and attempt to find the value. But suppose, instead, that they asked for $R(6,6)$, we should attempt to destroy the aliens.

Although mathematicians are believed to be interested in precision, they'll often settle for less. The fact that $R(5,5)$ is known to lie between 43 and 49 is almost as good as knowing that it's 46. (If this turns out to be the correct answer, I will claim credit for this major discovery.) But this tolerance for imprecision goes to ridiculous lengths with one particular Ramsey problem.

As well as dealing with subgraphs contained in complete graphs drawn on a plane, like the examples above, Ramsey theory can ask similar questions about subgraphs contained in three-dimensional complete graphs. Take the eight points at the corner of a cube, for example, and join each of them to every other point, and you get a nest of lines among which are various simpler shapes. You can then ask questions about subgraphs in that complete graph, including those that lie on a plane. All triangles will, of course, but subgraphs of four points or more don't have to.

For Ramsey theorists, two and three dimensions are the nursery slopes. The Ramsey problem with the world's most imprecise answer deals with complete graphs of higher dimensions. I will frame the question without attempting to explain it. (You need to know that a hypercube is the higher-dimension equivalent of a square in two dimensions or a cube in three.) What is the smallest dimension of a hypercube such that if all the lines joining all pairs of corners are two-coloured it will necessarily contain a complete graph of four points in one plane?

No one has yet found the answer to this, but the mathematician Ronald Graham has proved an upper bound for the answer. Like 49 for R(5,5), an upper bound is a figure that you can prove is equal to or more than the correct answer.

Graham's upper bound for the answer is so large that it needs a special notation to write it. And even writing it in a special notation would take up too much space here. Suffice it to say that this number is so big that if all the material in the universe were turned into pen and ink it would not be enough to write the number down in decimal notation.

The twist in the tail of this particular story is that it has been calculated in the last few years that the correct answer may be as small as 11.

'KNICKNACK GOOGLEWHACK' OR HOW GOOGLE WORKS

Most of us use computers without thinking much about how they work, rather in the way we drive cars without being too aware of what goes on under the bonnet. Facts about data handling and memory size in our computers may come to our attention when we are buying a new computer, but we are only dimly aware of the true level of achievement of this technology, which has become such an essential part of the modern world.

For me, the best illustration of the extraordinary feats computers are capable of comes every time I type a word or phrase into the Google search engine. When I type the word 'type', for example, within 0.16 of a second (the time is indicated on the screen) I receive the first page of a list of about 2,780,000,000 web pages that contain the word 'type'. That's information about nearly three billion pages, retrieved in less than a fifth of a second. If I type the words 'movable type', in 0.20 of a second I'm told that there are about 15,100,000 pages containing that phrase. And if I type the phrase 'the phrase "movable type"', the result is returned in 0.08 of a second, telling me that there are precisely eight web pages containing that phrase. Or rather, there are eight different web pages, because Google also tells me that there are a number of duplicates of those eight pages, so that the total number is forty.

What's going on here? Can it really be that a computer somewhere receives my request and then reads the entire contents of the Internet and collects the pages I need in a fraction of a second?

Actually no. What Google does is cleverer than that, though equally amazing. Google is continuously gathering web pages as they are created and adding them to its database. Each time it acquires a page, it creates a list of all the words on that page and adds those words to an alphabetical index, with a unique address by each word which indicates the page containing the word. So, to describe it in a very simplified way, the word 'type' in this index will have attached to it 2,780,000,000 or so page numbers. That entry with its list exists before you ever search for it, so the 0.16 of a second is merely the time taken to tell you something the computer already 'knows'. Higher in the index will be the word 'movable', with about 25 million page numbers. If I were to type in the word 'movable' and the word 'type' separately, i.e. not in quotes, Google would compare the two lists, of 2,780,000,000 and 25,000,000 page addresses, and make a separate list containing only the addresses that are on both lists, i.e. only the pages that contain both words. But I then put the words 'movable type' in quotes, meaning that I wanted only those pages that have the two words together, 'movable' followed immediately by 'type'. This is where a second piece of information gathered at the indexing stage comes into play. As well as storing the fact that 'movable' is in document 12, say, the index will also store the position of the word in that document, at position 31. So you can imagine a series of entries of the form (D12,31) for the word 'movable' in the index, containing the document number and the position. The index entry for 'type' might contain the reference (D12,32). From comparing the index entries, Google would know that the phrase 'movable type' is contained in document D12 with the two words at position 31 and 32 and it would include D12's address in the list it shows when you search for the phrase.

People with too much time on their hands have invented a game using Google's indexing system, called Googlewhacking. The game is to find a pair of words that occur on only one page in Google's vast archive. Words like 'onetiming' and 'lemming', for example, which appear only in a message board about hockey. You might think, since Googlewhackers have their own website where they list their discoveries, that the moment a new Googlewhack is listed it will no longer be a Googlewhack, since it will now be on two sites, its original one and the Googlewhack site itself. But Google has graciously excluded the page of new Googlewhacks from its indexing process, thus avoiding such a paradox.

MY INFINITY IS BIGGER THAN YOUR INFINITY

Most of us have a difficult enough time understanding the concept of infinity without having to worry about how there can be infinities of different sizes. But the fact is that mathematicians deal with several different 'sizes' of infinity, each of which is 'infinitely' greater than a lower one. Most of us think of infinity as the number you get to if you start counting from the number 1 and go on for ever. It seems nonsense to talk about numbers that might be bigger than this infinity (perhaps counting beyond for ever). One way mathematicians have tried to show that there are such numbers has been to use an idea called 'bijection', which just means one-to-one matching.

Supposing you lined up all the counting numbers (1, 2, 3, . . .) all the way to 'infinity' (that's the last use of quotation marks round the word – you'll have to accept that when I use the word in a way that suggests it's a number I know that it really isn't). If you had another set of numbers, the fractions, say, and you could pair the two sets of numbers so that for every whole number there was a fraction and for every fraction there was a whole number, all the way up to infinity, then you could say that each set contains the same number of members and therefore their infinities are equal.

Conversely, if you had a set of numbers that could not be paired, one to one, with the whole numbers without leaving lots of them over, unpaired with anything, then you'd have to say that the infinity that applied to the complete set of those numbers was bigger than the infinity of the whole numbers.

Take the fractions first. It may seem unlikely that there are only as many fractions as there are whole numbers. After all, between any two whole numbers – 1 and 2, say – there must be lots of fractions – $3/2$, $4/3$, $6/5$, and so on. But if I can arrange all the fractions in a unique order in an infinitely long list, so that the whole number 817, say, can be paired with the fraction at the 817th place on the list, every fraction will be paired with a unique number and every number with a unique fraction. (The whole numbers will be part of this list, because we can express 4, say, as $4/1$.)

Here's how to arrange that list. With each fraction, you add together the top and the bottom, and then arrange them in ascending order of each fraction's result, which we'll call s. (If the fraction has a negative number on top, you just ignore the negative sign.) So for 1/2, s equals 3; for 2/3 s equals 5; for 11/17, s equals 28, and so on. Some of these fractions will give the same s but since our only aim is to produce a long ordered list, we can invent a rule that tells us unambiguously which order to put them in. That rule could be that if several fractions produce the same s, we put them in ascending order of the number on the bottom of the fraction. So for seven fractions – −4/1, 1/4, 2/3, 3/2, 4/1, −3/2, −2/3−s equals 5. To order them by the number on the bottom we arrange them as: 4/1, −4/1, 3/2, −3/2, 2/3, −2/3, 1/4. Now we number each item in this long list of transformed fractions in sequence, so that each fraction is paired with one of the whole numbers, all the way up to infinity.

Every fraction will be represented in one place only on the list, matched with a whole number that shows its position on the list. No fractions are missed out, and no whole number is without a fraction associated with it, so there must be the same number of each when an infinite number of them are gathered together.

OK, so maybe we can accept that effectively all infinite collections of things have the same number of things even though it seems as if they shouldn't in the case of the fractions. But how could there be infinitely large collections of things that are *larger* than the infinity of counting numbers?

The German mathematician Georg Cantor found two sets of numbers that cannot be placed in a one-to-one correspondence with each other, in the way we did with the counting numbers and the fractions, above. He did it by starting with the assumption that they *can,* and finding a contradiction. If you had a theory that all swans are white, the discovery of one black swan would knock that theory on the head [🐕 25].

One of the sets of numbers Cantor compared was the natural, or whole, numbers that we used above. The other was what are called real numbers. The real numbers are equivalent to points on a line from o to infinity. So they include the whole numbers and the fractions, but they also include what are called irrational numbers, which cannot be expressed as fractions or ratios of whole numbers (ir-*ratio*-nal) [🐕 53], but can be expressed as a continuous line of decimal digits. Fractions can be expressed as decimals as well, but after a certain number of digits the rest are zero. So 5/8 is the same as .62500000000. . . whereas the digits in the irrational number 17.38279462900962835687648. . . go on for ever.

To prove that the real numbers cannot be placed in a one-to-one correspondence with the whole numbers, Cantor showed that, however you arrange the real numbers in an ordered list, in the way we did with the fractions, you will always be able to come up with a real number that is not on that list.

He reasoned as follows. Suppose we have a list of all the real numbers (an infinite number) and we devise a rule to arrange them in order. We might come up with a list that looks like the following:

Whole number	Real number
1	7.2728654901088. . .
2	2.0709903829756. . .
3	18.696243576675. . .
4	0.8717454638892. . .
5	3,834.2020203020. . .
6	0.6766682920082. . .
7	3.1416269873562. . .

Whatever the rule might be for these numbers, it's pretty impenetrable, but that doesn't matter. As long as we think we can assign every real number to one of the whole numbers we know and love, we'll end up with a list a bit like this.

Now, you might show me that list and say that you have a rule that means that every real number anyone could imagine is on it somewhere, all the way up to infinity, and therefore that the infinity of the real numbers is the same as the infinity of the whole numbers they are associated with. But whatever your list looks like, I can create a real number that is *not* on the list.

For simplicity we'll just look at the decimal part of the numbers, since it doesn't alter the argument.

I will construct a number whose first digit after the decimal point is different from the first digit in the first number on the list. The second digit of my number is different from the second digit of the second number on the list, the third digit in my number is different from the third digit after the decimal point in the third number in the list, and so on.

With the list I've given as an example above, I could create the number 0.3942501... The dots mean that it would have an infinite number of digits, as most of the real numbers do. Now I can prove that this number is nowhere on your list, whatever rule you have devised. Because of the way I have constructed it, it cannot be anywhere on the list, because it differs by at least one digit from every single real number on the list. It's a 'black swan' that proves that the initial assumption – that you have produced a one-to-one correspondence between all real numbers and all whole numbers – is wrong. The two infinities – the infinity of real numbers and the infinity of whole numbers – are fundamentally different, in a way that Cantor went on to develop into a whole new branch of the theory of numbers. Finally, it probably won't surprise you to learn that mathematicians think there are many more than two 'sizes' of infinity. In fact, there is an infinite number, and, to crown it all, this infinity is larger than any of the infinities it enumerates.

SURF AND SERVE

Over the last few years I have helped with the computing side of a number of scientific projects. They include the search for extraterrestrial intelligence, the hunt for very large prime numbers, and tests of algorithms for predicting the three-dimensional shape of protein molecules from their linear formula.

The reason I have been asked to help with such a wide range of important scientific research topics has less to do with my innate talents and abilities, and more to do with the fact that I own a personal computer.

The scientists running these projects, and a dozen or so others, are exploiting a hidden resource consisting of under-used computing time, amounting to millions of hours, which can augment their own computers in carrying out massive number-crunching tasks. For much of the time, even when we are sitting at them, our personal computers are lying idle. One of the earliest projects to make use of this time is called SETI, standing for 'Search for Extraterrestrial Intelligence', which gathers a huge amount of celestial radio data every day from a device attached to a large radio telescope on the island of Puerto Rico. The data appear to be a kind of 'white noise' emitted randomly by stars and galaxies, but the hope is that one day among the noise will be a signal from an extraterrestrial civilization, which will be identifiable by some regularity rather than randomness. By downloading a simple piece of software, home computer users can help to analyse these data, which are sent in regular bursts to each individual participant. In fact, if you share in this task you can watch your computer as the software carries out its analyses, and imagine the moment when your computer spots a regular pattern and alerts the world by sending a message to SETI.

It was a clever idea and has been taken up by other scientists with a large amount of data to process which don't require more sophisticated software than the average home computer.

Until now, you have had to apply to a project and download some software to take part. But I came across an ingenious use of your and my computer that doesn't even require our consent or knowledge. You may have noticed that sometimes when interacting with a website you are asked to read off the screen a distorted word or set of letters. This is done to ensure that there is a human being trying to use the site rather than a computer programme seeking to abuse online services in some way, for example to buy large numbers of concert tickets for resale at inflated prices. These words or groups of letters are called CAPTCHA, standing for Completely Automated Public Turing test to tell Computers and Humans Apart.

A new use for CAPTCHAs has emerged as a result of projects to digitize books and make their text available on the Internet. In the past this has been a labour-intensive task, using humans to read the books and transcribe them. More recently, a more cost-effective way has been to use optical character recognition (OCR) software, which reads books automatically at high speed and turns them into word-processible text. But the older the book, the more likely it is that the computer will have trouble reading it. A Victorian novel in tiny print on yellowing crinkled paper is going to present reading challenges to a computer that will cause no problem to a human reader.

This is where CAPTCHAs come in. Every day, humans type more than a hundred million of them, as they access websites. Computer scientists at the Carnegie Mellon University in Pittsburgh have shown how to use all this free effort by persuading some websites to use words as CAPTCHAs that computers have been unable to read while digitizing old books. In this way, with suitable safeguards to make sure that the readings are accurate by using the same word on different sites, they have created a system that does a job, the processing of unclear words, which would previously have required professional manual transcribers. In comparing their system, called reCAPTCHA, with computer optical recognition, the scientists report an accuracy rate of 99.1 per cent, compared with 83.5 per cent for standard OCR. After a year of this project, Internet users had unintentionally deciphered nearly five hundred million words, which is equivalent to the indecipherable words from over 17,600 books.

So next time you're presented with a distorted and semi-legible word on a website, try your hardest to read it, since you'll be adding to the corpus of world literature on the Internet, as well as buying yourself a ticket for the next Gilad Atzmon concert.

(W)RAPPING WITH MOZART

Since 1890, the burghers of Salzburg have feasted on *Mozartkugeln* – 'Mozart balls' in English – a sweet with a core of pistachio marzipan surrounded by layer of nougat and dark chocolate. Traditionally, these spheres are wrapped in square or rectangular silver-foil wrappers, which means that some of the foil is wasted in the creases that have to be tolerated in order to press the flat foil around the spherical chocolate.

In that ceaseless quest for new knowledge that mathematicians are well known for, a father-and-son team at New York University decided to work out the minimum size of silver foil needed to wrap a Mozartkugel. A significantly smaller wrapper would mean savings in the sweet manufacturer's silver-foil bills.

Two types of wrappers are currently used – one square, with a side length of π times the square root of 2, the other a rectangle with sides π and 2π. (It's nice to discover that even before the US mathematicians got their hands on the Mozartkugeln, a mathematical mind had been behind the design of the wrappers.) In each case, the wrapper's area was nearly 60 per cent greater than the area of the surface of the sweet, wasting about a third of the silver foil.

No doubt after unwrapping (and probably consuming) many Mozartkugeln, the mathematicians now believe they have found a way of wrapping the balls with less tinfoil. They have worked out that an equilateral triangle with a side of just under five times the radius of the balls could entirely cover the sweet and use 0.1 per cent less silver foil than the current wrappers. To those who might feel that this was a frivolous task for their undoubted talents, they argue tongue (and Mozartkugel) in cheek that their discovery could lead to a reduction in the carbon footprint of Mozartkugeln manufacture and 'partially solve the global warming problem'.

If I tell you that the company that makes authentic Mozartkugeln (there are several imitators) makes 1.4 million a year, you might be able to solve the following Fermi question [🐕 272]:

What weight of foil would be saved in a year if the factory switched to equilateral triangles for its wrappers?

THE FIREMAN HYPOTHESIS

The English mathematician G. H. Hardy, who worked in the abstract reaches of pure mathematics, tried in his book *A Mathematician's Apology* to counteract the idea that maths was a rarefied pursuit, of interest only to a small proportion of the population. Mind you he didn't try very hard – in one of his articles about mathematics he wrote that 'Landau's *Vorlesungen* [lectures on number theory] or Dickson's *History*, six great volumes of overwhelming erudition, [are] better than the foot-ball reports for light breakfast table reading'.

Hardy pointed out that many people enjoy chess or bridge, which both involve mathematical reasoning, and others enjoy the puzzle pages of newspapers. If he had been writing today he would also have pointed to the popularity of Sudoku puzzles.

A touching example of the way numbers can fascinate non-mathematicians came to light in 2007. (You'll see why it is touching later.) A New York fireman called Bobby Beddia told a friend that there was something special about his last birthday for him – he had reached what he called his 'birth year'. By this, he meant that it was the year in which he reached the age that was the same as the last two digits of the year he was born. Beddia was born in 1953 and he was therefore fifty-three in 2006. Everybody can work out his or her own 'birth year' – mine was 1984. The only people whose birth year is rather disappointing are those born in 1900 or 2000.

It turns out that for any one year (apart from 2000) there are two age groups who can claim it as their birth year, separated by fifty years. So in 2006, as well as fifty-three-year-olds like Beddia, there were three-year-olds, born in 2003, for whom 2006 was the year they became three.

Like many aspects of number theory, the Beddian year, as one mathematician has called it, starts with a simple observation but actually leads into some interesting questions that are not always easy to answer. While it's easy to work out your Beddian year from the year you were born, it's a little less obvious to work out in what year people were born whose Beddian year will be, say, 2014. An American mathematician, Barry Cipra, dug a bit more deeply and tried to work out, for any one year, what the age range is of people who are pre-Beddian, i.e. had not yet reached their Beddian year. He worked out that there were actually two age ranges involved. Taking the year 2007, Cipra discovered that children aged nought to three and older people aged between eight and fifty-three would not yet have reached their Beddian year in 2007. The rest – four to seven-year-olds and fifty-three to ninety-nine-year-olds – would all have passed their Beddian years. It doesn't require advanced mathematical techniques, but it still needs a bit of mental juggling, with two sets of numbers, years and ages, and with the fact that some lives straddle the end of the century.

Cipra may well have exhausted the deeper possibilities of Beddian theory, but he was himself surprised at how such a simple observation led to a tricky problem or two. Unfortunately, Bobby Beddia himself never read what a mathematician had made of his discovery. A month before the end of his Beddian year, he was killed attending a fire in a vacant office building near the site of the World Trade Center terrorist attack.

WHAT A COINCIDENCE

The mathematician Jack Littlewood was once asked about the most remarkable coincidence he had experienced. He wrote:

> A girl was walking along Walton St. (London) to visit her sister, Florence Rose Dalton, in service at number 42. She passed number 40 and arrived at 42, where a Florence Rose Dalton was cook (but absent for a fortnight's holiday, deputised for by her sister). But the house was 42 Ovington Sq. (the exit of the Square narrows to road width), 42 Walton St. being the house next further on. (I was staying at the Ovington Sq. house and heard of the occurrence the same evening.)

Many of us have experienced or come across such events and it is difficult not to believe that there is some deeper significance in them. And yet, our surprise is often a result of not hearing the whole story, or of ignorance of the science of probability.

To take the first, if someone calls you on the phone and correctly names the winner of a horse race that is due to take place, then rings you again the following week and again correctly names another winner, you might be tempted to take up his offer to buy the name of a third winner in a race the following week. But supposing I told you the following: that this man had rung a hundred people before the first race, with ten runners, and named each of the ten horses to ten people, and then phoned the ten who'd been given the first winner and told each of them the name of a horse in the second race. One of that hundred – you, in this case – would have been told of two winners in succession and would be very tempted to buy a third name, even though it would turn out that you had only a one in ten chance of your horse winning the third race.

One often-told 'spooky' coincidence centred on a novel written in 1898, called *The Wreck of the Titan*, which featured a ship called the *Titan* which sank on its maiden voyage in April, as a result of a collision with an iceberg. Fourteen years later, the *Titanic* sank on its maiden voyage in April after colliding with an iceberg, and 1,500 people died, partly as a result of there not being enough lifeboats. In the wreck of the *Titan*, nearly three thousand died.

This coincidence was much more likely than it might seem. Assuming you wanted to write a dramatic novel about a ship-wreck in 1898, you would need a name, a route, a cause of shipwreck and some factors – such as a large number of casualties and a high-profile voyage – to give it more drama than a routine shipwreck. Almost the exact details that applied in the case of the *Titanic* fall into place with the choice of topic. To start with, the ship should be big, and therefore have a name that resonates with its size. *Gargantua, Giant, Colossal* and *Huge* are clumsy names for ships, but something mythical like – oh, I don't know – *Titan* might work. If it was to be a big ship, with English and American passengers (the author was writing for the English-speaking market), then you wouldn't choose a trans-Pacific or trans-Indian Ocean route, so the Atlantic is the likely choice. What is the most likely cause of shipwrecks in the Atlantic? Icebergs. And what time of year are icebergs particularly hazardous? April.

Often when we hear or read about amazing coincidences the version we are told, for dramatic effect or in order to deceive, is embroidered with untruths. Here's a story from a book published by Reader's Digest, called *Mysteries of the Unexplained*:

In Detroit in the 1930s, a man named Joseph Figlock was taking a walk down a residential street when a baby fell out of a second-storey window onto Figlock. Figlock caught the baby and both he and the baby were no worse for the wear. One year from that date, Figlock was again walking down the street when the same baby fell from the same window onto him. Once again, he returned the unharmed baby to its mother and Figlock, who was also unharmed for the second time, continued on his way.

Pretty extraordinary, if true. But it isn't. If we read *Time* magazine for 17 October 1938, we find this:

Coincidence In Detroit: Street Sweeper Joseph Figlock was furbishing up an alley when a baby plopped down from a fourth-story window, struck him on the head and shoulders, injured Joseph Figlock and itself but was not killed. Last fortnight, as Joseph Figlock was sweeping out another alley, two-year-old David Thomas fell from a fourth-story window, landed on ubiquitous Mr. Figlock with the same results.

Not the same baby and *not* the same window – two elements in the story that made it more incredible. It was still an unusual experience, no doubt, but even Mr Figlock's occupation was one where, since he spent his entire working day in the streets and alleys of Chicago, if babies were going to fall on anyone, they'd be more likely to fall on him than, say, a telephone operator who was indoors all day.

The effect of a personally experienced coincidence can be very powerful, particularly on people who are unfamiliar with statistics. Martin Gardner gives a rough-and-ready example of how stories of precognitive dreams can come about:

> Assume a woman dreams that her Aunt Mary dies in a fire. In the same dream Aunt Mary's husband escapes the fire by jumping out a window and breaks a leg. A few days later one of the following events takes place: Aunt Mary dies of an illness, her husband breaks his arm in an auto accident, or a house in the neighbourhood catches fire. If Aunt Mary dies the dreamer will be able to tell friends that only a few days ago she dreamed that her aunt died. If the husband breaks an arm, the dreamer may recall that in her dream he broke one of his bones, she isn't sure which but she *thinks* it was his arm. And of course if a house nearby catches fire she will recall that aspect of the dream. Other events in the same dream, of which there could be scores, will be totally unremembered.

I was once in the library of Trinity College, Cambridge, and happened to pick a book off the shelves which had a photograph of a vaudeville performer, Mrs Senrab. I showed this to a colleague who was with me, and asked whether he saw anything strange about her name. He said 'no', and I pointed out that it was actually 'Barnes' backwards. Whether she had just turned her own name backwards to produce a more exotic stage name I don't know. I then said, 'It's like the street in Washington called "Tunlaw" which is actually "Walnut" backwards.' There was a gasp from a nearby table, where a young American woman was working. 'I *live* on Tunlaw!' she shrieked, quietly. What caps the whole story is that my colleague lives in Barnes.

There are two questions that arise from coincidences of this sort. One is, if they are not merely the result of concatenations of chance events, what purpose do they serve? For precognitive dreams of disasters to be significant, they should, perhaps, help to prevent the disaster, but I have come across no evidence of a dream of an air crash, say, that led to inspection of a plane and discovery of a fault in an engine. For other less dramatic incidents, such as my library experience, there was no conceivable usefulness to that odd set of events. For instance, the lady from Tunlaw Street wasn't revealed as a long-lost relative (although my maternal grandfather did later turn out to have lived on *Chestnut* Street in Harrisburg, PA).

The second question is: supposing we lived in a world where there were no such coincidences? Would anyone notice? Well, the answer is: yes. Scientists and mathematicians would be extremely puzzled by the fact that, in spite of every probability calculation suggesting that such events should happen regularly, there was an eerie absence of precognitive dreams or novels that had vague similarities with subsequent events. That *absence* would be far more worthy of explanation than the *existence* of such events is today.

NAMES THAT COUNT

The names of numbers are so familiar to us that it's difficult to imagine what it was first like when people came upon the whole idea of counting. The anthropologist Alfred Gell described the reaction of children of the Umeda tribe in Papua New Guinea when they first learned how to count:

I was present during one delirious afternoon when the children finally did catch on to the basic principles of number – the fact that with numbers you can count *anything*. Released from the schoolhouse, the excited children ran hither and thither in little groups, applying their new-found insight: they counted the posts of the houses, the dogs, the trees, fingers and toes, each other – and the numbers worked, every time.

Our own number names are based on the decimal system. Most of us are aware of how number names change from ten to the teens and from a hundred onward, and how that relates to the basis of the system being the number ten. When anthropologists started exploring the mental capacities and skills of different people around the world, they found that although 10 is the predominant base of numbers in the developed world, there was a bewildering range of alternatives used by people who had had to devise systems for themselves.

Some people use two as a base. The Gumulgal of Australia, the Bakairi of South America and the Bushmen of South Africa have a word for 1, and another for 2, and they build on those. So when the Gumulgal count from 1, it's as if they were saying 'one; two; one-two; two-two; two-two-one' and so on.

The most common grouping after 2 is 5. The Zuni of North America have an eloquent system of numbers based on the use of the fingers of one hand. Translated, their number words are:

1. 'taken to start with'
2. 'put down together with'
3. 'the equally dividing finger'
4. 'all the fingers all but done with'
5. 'the notched off' (presumably the thumb).

But not all 5-based systems use fingers. The Abipones of South America use *geyenknate*, 'emu's toes', for 4 and *neenhalek*, 'a beautiful skin with five colours', for 5.

With a small base, like 2 or 5 or 10, although the numbers can be clumsy, it's easy to work out how to count as far as you like. Another popular base is 20, the number of fingers and toes. Among the Galibi of Brazil, 20 is '*poupou patoret oupoume*', meaning 'feet and hands'. But sometimes, anthropologists have come across much more cumbersome bases. The Kewa of New Guinea use the base 47 for their number system. They have arrived at this by counting round the body – starting with fingers and thumb, the heel of the thumb, and the palm, on one side, and so on via the arms, neck, eye, between the eyes, and then back to the other side of the body. This system has the advantage that by pointing to any of these parts of the body you can quickly convey a number between 1 and 47.

It's difficult to imagine how the anthropologists who have gathered these fascinating data are seen by the people they interrogate, but we get occasional glimpses.

A French anthropologist called Houton La Billardière was on a voyage to Tonga when the expedition's ship was beached and he had to spend an extended period with the Tongans. He used his time to good effect. He was interested in the words the people used for large numbers and found that the Tongans had an extensive vocabulary, which he carefully wrote down. He was surprised to find that they had specific words for numbers, such as 10,000,000, which they said was *laoalai*, and 10,000,000,000, which was *tolo tafai*. He explained this facility with large numbers as follows: 'We ought to reflect, that a people who are in the frequent habit of counting out yams, &c. to the amount of one, two, or three thousand, must become tolerable good numerators, by finding out some method of rendering the task of counting more easy.'

But a later anthropologist who understood Tongan better pointed out that the Tongans had been playing games with La Billardière. The word they gave for 10,000,000 meant 'foreskin'; for 10,000,000,000 'penis'; and other words had equally playful meanings, finishing up with the biggest number La Billardière could think of, which the Tongans told him was called *ky ma ow*. It was later discovered that this meant 'eat up all the things we've just told you about'.

Oh, how they must have laughed around the cooking pot that evening.

FLORA AND FAUNA

ONE LEAF OR TWO?

In the church at Grantchester near Cambridge there is a mosaic floor surrounding the altar. Regularly spaced over this floor are alternating flower designs. One is a six-petalled lily and the other a five-petalled rose, emblems of Corpus Christi College, Cambridge, patron of Grantchester church. You might easily pass it by without a glance – just two flowers, chosen by the artist for no apparent reason. But coincidentally, these two flowers and the numbers of their petals represent one of the most fundamental divisions of flowering plants, a division that is crucial for plant classification.

Unlike many ways of classifying the natural world, this particular division was discovered late. When you consider the millennia during which people have grown, exploited and contemplated plants, this particular division was not noticed or described until the seventeenth century, by the scientist John Ray. He observed that the seedlings of flowering plants differed in subtle but fundamental ways. In a paper contributed to the embryo Royal Society he reported that 'the greatest number of plants spring out of the Earth with two leaves, for the most part of a different figure from the succeeding leaves . . . the seed-leaves are nothing else but the two lobes of the seed'. He then went on to write of a smaller group of plants, those whose 'seeds spring out of the Earth with leaves like the succeeding [ones] . . . nor have their pulp divided into lobes'.

The two different types of seedling that Ray noticed are called 'monocotyledon' for the seeds that produce a single seedling leaf and 'dicotyledon', for paired seedling leaves. Nowadays, for working botanists discussing the plant world, the names 'monocot' and 'dicot' trip off the tongue. The 'cot' is short for cotyledon, the seedling leaf. Gardeners will know that onions produce a single long leaf while cabbages produce a pair of seed leaves, and so a germinating onion seed is a monocot and a germinating cabbage seed is a dicot.

In a way, it's not surprising that this fundamental division escaped detection for so long. There are about 250 families of dicots and fifty or so monocots, and if you look at the adult plants that make up each group, they seem to have very little in common. Monocots include lilies, rushes, sedges, grasses, irises, orchids and palm trees. Dicots include honeysuckles, sunflowers, buttercups, roses, mustards, mallows, primroses, phloxes, snapdragons, mints and geraniums. The characteristics that separate out the two groups are the sort of thing you notice only if you get down on your knees and peer at the different components of the plant, preferably with a magnifying glass.

But in fact, the two groups have very different roles in the lives of humans. One botanist has pointed out that 'the wind-pollinated monocots are the things by which we live'. The bread, porridge and rice we eat and the beer we drink are all derived from monocots, and the fact that the cereals from which these foods come are wind-pollinated means that they look insignificant to us, compared with dramatically flowered plants [🐕 97], because they don't need to attract pollinators to propagate.

Roots, leaves, flower parts, even the layout of tiny tubes that make up the plant's vascular system, all are linked to this earliest visible sign of difference – one leaf or two at the seedling stage. And there's one other easily visible but long unnoticed characteristic that separates the two groups. The flowers of monocots have three or six petals and dicots have four or five. So the lily in the Grantchester mosaic with its six petals is a monocot and the rose with five is a dicot.

FLOWER POWER

Most of us have a rough idea of the relationship between flowers and some insects and birds. The flowers on plants act to persuade creatures such as insects or small birds to alight on them, so that the pollen – containing the plant's sperm cells, which fertilize other flowers – can stick to the visitors' legs or bodies and be carried to the female organs of the other plants they visit. The job of flowers is to attract, with a combination of bright visual displays and seductive scents from its nectar. But different plants use different strategies.

Results of an investigation carried out in 2007 showed the subtlety of the strategies used by different plants to maximize pollination. A group of scientists from Cornell University were puzzled as to why one particular plant, *Nicotiana attenuata*, laced its attractive nectar with nicotine, a chemical that repels the very creatures it is trying to attract. The main pollinators for *N. attenuata* are a hawkmoth and a hummingbird, both of which dislike the smell and taste of nicotine.

The researchers realized that as well as needing to attract certain creatures, *N. attenuata* also had to repel others. Certain caterpillars like to eat flowers, so they had to be kept away, and so did some bees, which rob the plants of nectar by drilling into the side of the flower without carrying away any pollen.

One final piece of the jigsaw was the need for flowers to be attractive but not too attractive. Some orchids, for example, have no nectar and rely on their complex beauty to attract pollinators. When researchers added nectar to these orchids, pollinating insects remained longer at each plant and therefore were unable to visit and pollinate so many other flowers. So by being *too* attractive a plant can reduce the effectiveness of later pollination. The longer an individual insect or bird spent at any one flower, the fewer other flowers it was able to visit to deposit the pollen it had gathered.

How did all these factors explain the properties of the flowers of *N. attenuata*? The researchers decided to manipulate the proportions of the two key substances, the attractive scent in the nectar and the repulsive nicotine, by breeding different versions of the plant with modified genes, ranging from flowers with no scent and lots of nicotine, to very scented plants with no nicotine. They set these plants out for the pollinators, and observed which insects and birds went to which flowers and for how long.

They discovered that the combination of scent and nicotine that exists in nature was exactly the right balance to achieve its aims. They described the plant as exhibiting 'push' and 'pull' strategies. The floral enemies were kept away while its pollinators were attracted. And the nicotine, in addition to repelling the enemies, was unpleasant enough to make sure the pollinators only stayed a short time and visited more plants in a day.

MY NEIGHBOUR, THE STICKY HAIRY PLATE

Until the last twenty or thirty years, deductions about which animals were related to which in the animal kingdom were based on the comparison of shapes, sizes and other visible characteristics of animals and their bones and organs, known as morphology. But the genetic revolution that began with an understanding of how DNA in genes carried instructions for making every element of a living organism has led to surprising discoveries about the relationships between different creatures. What 'relationship' means, in this context, is the extent to which animals are on the same or nearby branches of the tree of life, a diagram with a trunk and large branches and smaller twigs which is based on a combination of biology and the study of fossils.

Charles Darwin wrote:

> The green and budding twigs may represent existing species; and those produced during former years may represent the long succession of extinct species. At each period of growth all the growing twigs have tried to branch out on all sides, and to overtop and kill the surrounding twigs and branches, in the same manner as species and groups of species have at all times overmastered other species in the great battle for life.

The 'twigs' of the tree represent species that are still around on Earth today. Two species on adjacent twigs are closer together biologically – and usually more alike – than species on opposite sides of the tree.

In the last twenty or so years, the task of placing different species in their correct part of the tree has increasingly used the analysis of the genes of different species. Roughly speaking, the more alike the sets of genes of two species are, the closer together they are on the tree. In some cases, the results of this technique have overturned traditionally accepted relationships. When applied to birds, for example, grebes, which have been seen as related to loons, are now actually closer to flamingos; the nightjar, a dull brown bird, is now cousin to the iridescent hummingbird; and parrots and songbirds are closer than anyone ever realized.

Some researchers have used these techniques to trace the family relationships between humans and other creatures. There are various estimates for the amount of shared DNA between humans and chimpanzees, for example, some going as high as 99.4 per cent of the most critical genetic sites. It is perhaps not surprising that there is a close relationship between apes and humans. Such closeness was evident from the old morphology arguments. But a recent discovery using genetic techniques has revealed a much more surprising close relative, one that would never have been revealed through comparisons of shape or structure or shared anatomy or physiology.

Trichoplax adhaerens is a member of the animal kingdom whose genome – the full set of chromosomes – looks 'surprisingly like ours', according to Daniel Rokhsar, an evolutionary biologist at the University of California. Part of Rokhsar's surprise comes from the fact that *Trichoplax adhaerens* is about as unlike a human as any living organism could be. Its English name is 'sticky hairy plate', and if you were next to it in a zoo or aquarium not only would you not recognize it as a relative, you probably wouldn't see it at all.

Trichoplax is about a millimetre long, for a start. It is made up of only four types of cells, and it possesses none of the organs or systems that our genes construct in the developing human embryo – no stomach, muscles, nerves or gonads. Not even a head. It glides along like an amoeba and acquires its food by releasing digestive enzymes from cells on its surface which break up algae as it glides over them.

So what is it that makes it more like humans than anyone suspected?

Rokhsar and his colleagues worked out that sticky hairy plate has a sequence of 11,514 genes, containing many of the counterparts of the genes of much more sophisticated creatures like us. Genes that are needed to make body parts that *Trichoplax* doesn't have; genes to make proteins that the mammalian body needs to coordinate the specialized functions of different cells, which *Trichoplax* also doesn't have. Somehow, the genetic information that came to be an essential part of much more complex creatures was already in place in very early creatures like *Trichoplax*.

The genetic analysis has cleared up one other issue. Some biologists thought that *Trichoplax* was a representative of the oldest branch of the family tree, but in fact it turns out that it is younger than another candidate for that honour – the comb jelly. The members of this group have a more interesting appearance than sticky hairy plates, and include the sea gooseberry and Venus' girdle, a shimmering, iridescent creature that grows up to a metre and a half long.

With 'phytogenomics', as this technique is called, still in its infancy, it's possible that the tree of life is going to undergo much more pruning and grafting in the years to come.

HALF ANIMAL, HALF PLANT, AND GOOD ON TOAST

When you eat mushrooms *à la Grecque* or a truffle omelette, you may have assumed that you are eating a vegetable. But you aren't. Fungi (which can be pronounced 'funj-eye' rather than 'fun guy') were thought for many years to be plants, members of the vegetable kingdom with a few peculiarities. But modern science has shown that they are not.

More than 56,000 species have been described, from nasty diseases that form between our toes to the mushrooms that are good on toast, and since the advent of humans, fungi have made their mark on our history in a number of ways. Some believe that the legendary 'Pharaoh's Curse' was due to a long-lived fungus that lay dormant in Egyptian tombs until they were opened by archaeologists. Ergot, a fungus that contaminated rye in the Middle Ages, led to bouts of madness; in the nineteenth century, potato blight, another fungus, destroyed the lives of hundreds of thousand of Irish who starved to death or emigrated. And phylloxera, a fungus on grapes, dealt a devastating blow to the French wine industry.

But fungi have also contributed to the enjoyment of mankind. Bread would not rise without them; many fungi are edible in themselves; others turn sheep's milk into Roquefort cheese, or cottage cheese into Camembert. And one fungus, *Penicillium notatum*, was the source of the world's first and most widely used antibiotic.

These living organisms have evolved in the same environment as plants and animals and often carry out tasks that are essential to the survival of their neighbours. One fungus, for example, lives among certain grasses and monitors the health of the individual blades. If one blade of grass seems to be ailing in some way – short of water or food – the fungus will build a bridge between that blade and an adjacent healthy one, and pump over a fresh supply of water and food. Another fungus has a three-way relationship with certain trees and with squirrels. The fungus grows in the tree's roots and enables it to gather nutrients from the soil. At the same time, it produces mushrooms which are eaten by squirrels who spread the fungus to other trees via their faeces.

Scientists no longer rely on the *appearance* of a lifeform to fit it on to the tree of life. It was the development of sophisticated DNA analysis [🐕 199] which revealed the differences between fungi and plants. The cell walls of fungi don't use cellulose, as plants do; and the chemicals that do the work of digestion are different from those in plants.

The results of DNA analysis on fungi have shown that there are three biological kingdoms – animals, plants and fungi – which separated out from a common ancestor about a thousand million years ago. The fungi split in turn into at least eight distinct lines, from aquatic forms called chitrids to baker's yeast, penicillin, plant rust, mushrooms and toadstools. Like plants and animals, fungi come in a bewildering variety of forms, from the microscopic, like the yeasts, to the very large. The biggest recorded edible puffball, delicious sliced and cooked in butter, measured 2.6 metres in girth and was capable of producing several billion billion spores, the genetic material that can produce other puffballs. On this basis it has been calculated that the spores from one large puffball could populate several galaxies with its descendants.

One type of fungus, called *Armillaria ostoyae,* can grow to occupy 1,500 acres and weigh several hundred tonnes. An example in Washington State in the USA is believed to be 1,500 years old, and every cell of the fungus has identical DNA. Far larger than blue whales or sequoia trees, this fungus, as one single organism, is the largest living entity on Earth.

WHO INVENTED THE WHEEL?

Wheels seem a uniquely human invention. No creature has evolved to possess a set of wheels, which would be quite an effective means of locomotion in nature. It's sometimes said the salamander provides one natural example of a wheel because it curls up into the shape of a rubber tyre and rolls down hills. A type of caterpillar does a similar thing. But on that basis, the first people to use logs as rollers to move large stones could be said to have invented the wheel. But the key insight was to make an axle, so that the object being moved could travel long distances without the rollers having to be replaced or repositioned every few feet.

The true wheel in nature is difficult to imagine, since living parts of humans and animals require a blood supply. Even fingernails and hair, which seem fairly inert, need to be regularly nourished, and it's difficult to see how a blood supply could have evolved that didn't get irretrievably tangled as the flesh and bone wheel rotated.

But on a smaller scale than humans, salamanders and caterpillars, there is a life form that could lay claim to have invented – or at least evolved – a type of wheel. Some bacteria use long filaments to move around in a liquid environment. But they don't just wave around like oars on a rowing boat. They are actually capable of rotating along their length, with the root of the filament held loosely in a socket on the surface of the bacteria. It's as if the filament – or flagellum, as it's called – is the axle and the body of the bacterium is the wheel. Around the root is a collection of molecules which together act like a motor and can cause the flagellum to rotate several hundred times a second. In 2008, scientists at the University of Oxford discovered that such 'motors' even have a clutch, a molecule that engages or disengages with the flagellum so that at times when the bacterium needs to remain motionless, the flagellum doesn't rotate.

FLIPPING CRAYFISH

Actions like the almost instantaneous withdrawal of a hand from a flame are a vital protective mechanism for living creatures. Those animals that did not have the speed of action necessary to escape danger without thinking about it would have died, leaving as survivors those who had this faculty, which they passed on to their descendants. These actions are called reflexes and are very different from most of our activities as we go about our daily lives.

Take the sequence of events that make up our response to the sight of a glass of freshly squeezed lemonade – 1. 'Ah, I see a glass of lemonade.' 2. 'Do I feel thirsty?' 3. 'Yes, I do.' 4. 'I'd better order my arm and hand muscles to lift the glass to my mouth.' 5. 'Now, I'd better open my mouth and create the necessary suction to cause the lemonade to enter my mouth,' etc, etc. While, in theory, the sense message from a hot flame has something in common with the sight message from a glass of lemonade, our responses are very different. We *don't* say, 1. 'Ah, a searing hot flame.' 2. 'Would I like to avoid it?' 3. 'Yes, I would.' 4. 'I'd better order the muscles of my arm to remove my hand from the flame.' If we did go though that process, we'd probably end up with a few charred stumps instead of fingers.

We and other creatures have evolved actions that bypass the decision-making process to save time. In a reflex action, the path between the stimulus – the flame – and the response – the muscle movement – does not go via the brain, but on a much shorter journey through the spinal cord at a lower level. In fact, this doesn't mean we are not consciously aware of the event, but by the time we have registered the pain of the flame, the hand has already moved away.

One of the creatures that is often used to study nerve activity is the crayfish, which has an important reflex known as the backward tailflip, used by the creature to escape very quickly from some indication of imminent danger. The muscles that cause the 'flip' are triggered to contract by a signal from a giant nerve fibre in the crayfish's abdomen. Individual sense messages from some threat, such as a burst of water or the prod of a predator, converge via different nerve fibres on the giant nerve fibre, at junctions called synapses. And this fibre sends its message to the muscles when it receives a number of sense messages all arriving at the same time, rather like a sudden influx of 999 calls at a central police station suggesting that something really bad is happening.

For years, scientists were puzzled by one aspect of this reflex – how do all the sensory messages arrive at exactly the same time at the giant nerve fibre when they've been stimulated at different parts of the crayfish? If they arrived at different intervals the fibre would never fire, because different parts of the crayfish are always experiencing individual sensations of pressure, and it's only when the whole body experiences them simultaneously that it is likely to be a threat.

Why was this seen as a puzzle? Well, the crayfish has sense organs of very different lengths. Its two antennae are longer than its body and wave around in the water while its antennules, poking straight out from its head, are much shorter. What's more, messages can start at any point along the antennae, and at more than one point at the same time. This suggests that a cluster of alarm signals would arrive at different times at the synapses and fail to trigger the muscle action. Although the distances the signals have to travel might seem short, the whole reflex happens in a fiftieth of a second, and if some signals have to travel twice the distance of others, with different arrival times, the backflip won't be triggered.

Somehow, scientists knew, the timing of the signals was varied so that they all arrived at the same time, but how was this done?

In 2008, the mystery was solved. Measurements of nerve impulse conduction speeds showed that the impulses from the farthest tips of the antennae travel faster than impulses triggered nearer to the point of contact with the body. So if a single burst of water triggers nerves to fire simultaneously at different points in the antenna, the signals from the sensors nearer to the crayfish body will dawdle and wait for the farther signals to catch up, so they can all arrive at the same time.

This may sound a rather complicated process, but in fact it's achieved quite simply. The speed of travel of an impulse along a nerve fibre is related to the diameter of the fibre. In the crayfish, the nerve fibres that run along the antennae increase in diameter the nearer they are to the crayfish body, ensuring that, like shy teenagers arriving en masse at a party, all the messages arrive at the giant fibre at the same time and trigger the tailflip.

THE OLDEST LIVING ORGANISM ON EARTH

In 1964, a geologist was carrying out research in the White Mountains of California. He was taking core samples from a clump of ancient trees known as bristlecone pines. These trees were known to be thousands of years old, with a largely dead core but a living bark. The samples could be used to find out the age of a tree, as well as facts about the climate during the time a particular layer of bark was being formed, and the geologist was looking for data from which to infer the size of glaciers that had existed at earlier epochs.

Unfortunately, his coring device snapped, and the geologist asked permission of the US Forest Service to cut down the tree so that he could get the necessary data from the pattern of tree rings that was revealed. They agreed, and by doing so they killed what was at the time the oldest living organism on Earth. The tree turned out to be 4,950 years old. It was as if a doctor taking a biopsy to determine the state of a patient's health had sought permission to kill the patient instead in order to do a post-mortem, because his biopsy needle had broken.

The felled tree had a younger neighbour, discovered in the 1950s and still alive today, estimated to be 4,796 years old and so coming up for a 4,800th birthday round about 2012. These trees live a long time because they have a slow metabolism, with the living layer of bark growing only an inch or so every hundred years. The pine's needles last for thirty or forty years, and even cones from the oldest trees can produce seeds that will grow today.

Until 2008, the bristlecone pine was believed to be the oldest living organism in the world. Then a tree scientist announced the discovery of a clump of spruce trees in the mountains of Sweden, one of which was 9,550 years old. At a stroke, the age of the Earth's oldest living inhabitant had doubled.

This spruce tree began to grow at a time when agriculture was in its infancy, the earliest cities were beginning to form, and the wheel was invented, thousands of years before the rise of the principal ancient civilizations of the East and the Mediterranean.

As with the largest living organism, the *Armillaria* fungus [🐕 103], the uniqueness of the Swedish spruce trees was revealed by DNA analysis. Unlike the usual trees we are familiar with, these trees had many trunks. A single trunk might last for six hundred years, but as soon as the old one died a new trunk appeared near by, with identical DNA and therefore part of the same organism.

And the scientist who made this discovery had a very appropriate name, Leif – Professor Leif Kullman.

ANIMAL MAGNETISM

For years there have been experimental results suggesting that birds use the Earth's magnetic field to navigate. In the northern hemisphere, birds fly north in the summer to breed in the Arctic and south in the winter to warmer regions. But nobody has been sure how this ability works. Concentrations of magnetic particles have been found in bird skulls, but with no apparent connection with any sense organs, and so it's difficult to see how these could be part of a navigation system.

A group of scientists in Oxford and the USA recently came up with a theory that shows how ingenious nature can be in solving problems that improve survival. Their proposed system has nothing to do with magnets in the head but instead depends on short-lived molecules whose lifetime depends on the surrounding magnetic field.

The scientists suggested that there are molecules in the light receptors in a bird's retina that undergo a chemical reaction when they absorb light. This reaction creates showers of particles which survive for about a millionth of a second. But the *precise* survival time, and therefore the numbers still around after a certain interval, is affected by the Earth's magnetic field. Somehow, the scientists believed, a sensory system in the bird's retina monitors these molecules and how long they live, and uses the results to set its direction of flight.

There were two problems with this theory. First, no one has found any evidence for this system in the physiology of bird vision. And second, nobody had seen a molecule that behaved in this way. In 2008, the second problem – at least – was solved, when scientists reported that they had synthesized molecules in the laboratory which were similar to those used in bird vision, and which behaved in just this way in a magnetic field.

In the same year, another strange discovery about animal magnetism was announced. This time, no complicated laboratory quipment was needed. In fact, anyone could have made the discovery since all it involved was studying pictures on the Google Earth website.

German scientists who had shown that some small rodents are sensitive to the Earth's magnetic field wondered whether larger animals were also able to detect such fields. They considered the migration patterns of herds of cattle or deer – difficult to herd into the laboratory – and hit upon the idea of using satellite imagery, freely available on the Internet, to hunt for herds and work out the direction they were facing.

They surveyed images of more than ten thousand cattle around the world and found that the animals tended to face north or south. What's more, it was *magnetic* north or south, which varies slightly from geographic north and south, defined by the position of the north and south poles. One of the problems with scientific experiments, particularly when you know what you are looking for, is what's called observer bias. However hard you try to be objective, if you are trying to judge what direction a particular herd is facing, particularly when you know which direction you'd *like* them to be facing, it's easy to introduce bias into the results. To try to remove this factor, the scientists assigned university students to analyse the same set of images and they came up with the same result.

No one knows why cattle and deer prefer to face north or south, but the scientists pointed out that if this is a general preference, it might be worth taking it into account when building cowsheds. If contented cows produce more milk, dis-contented cows forced to face east–west might produce less.

THE BIRD THAT KNOWS PHYSICS

Life on Earth is, by definition, biology-centred. The mechanisms and processes that keep creatures alive are the stuff of biology. But we live in a world of physics, too. Our biology has to recognize and find ways of dealing with the laws of the physical world. An elephant's legs are thicker than mine because they have to prevent an elephant's very heavy body crashing to the ground under the force of gravity. A mayfly has no such needs. Its concerns have little to do with gravity and much to do with the forces of wind and air, which determine how it can move in its world.

These physical facts often determine the shape that a creature has evolved to make best use of the opportunities presented by its environment. But they can also determine how a creature behaves, and reflect an apparent understanding of the physics of the everyday world.

The red-necked phalarope is a shore bird with a very long, thin beak, which feeds on tiny crustaceans. In two very interesting ways, it uses the physics of its watery environment to gather its food. These birds can often be seen rotating in tight circular paths in the water, dipping their beaks about once a second into the water. They are creating a vortex under the water that disturbs the river or lake bed and sucks its food nearer the surface, where it can be picked out.

Some other water birds, having gathered a small volume of water that contains the prey, will filter the food out by sucking the water through a kind of mesh that captures the food. But phalaropes appear to 'peck' the food out of the water and hold it in droplets in the tip of their long beaks, which are like needle-nosed tweezers. For some time, scientists were puzzled about how the birds moved the prey all the way along the beak to the throat, where it can be swallowed. Some birds jerk their heads back and use the inertia of the prey to 'throw' it to the back of the throat. But the phalarope's food is too light for this to work. In fact, it seems that the bird deliberately chooses crustaceans that do not exceed a certain mass, even though the fatter the prey the more satisfying it would be.

Somehow, a droplet of water containing a crustacean is moved from the tip of the beak to the bird's throat where it can be swallowed. Some long-beaked birds achieve this kind of movement with suction or by using their tongues. But the phalarope does neither. Instead, it relies on surface tension, the force that occurs at the surface of a liquid that makes it form a drop when placed on a solid surface.

Surface tension between a raindrop and a glass windowpane means that if the drop is not too large the forces round the edge of the drop will hold it in contact with the glass. In the same way, a drop of water containing a crustacean in the beak of a phalarope will be 'stuck' between the surfaces of the upper and lower jaw of the beak, if it is not opened too wide. To move the drop along the bill, the bird rapidly opens and closes its beak very slightly. First the drop spreads out, as the end nearest the throat moves back with the slight opening of the bill, then the end nearest the tip follows and closes up the droplet as the bill closes again. By what has been called a 'ratcheting' effect, the drop containing the prey is transported rapidly up the beak, at speeds of up to a metre a second.

In this interplay between the evolution of the phalarope and the physics of water in contact with surfaces, several things have had to evolve to be at their optimum for this particular form of feeding. The shape of the upper and lower beak surfaces, the physical nature of those surfaces enabling them to have just the right 'wettability' to match the needs of the droplets to move, the patterns of beak movement that the bird uses to achieve the movement of the droplet, the innate sense in the bird of what size prey will be light enough to move under surface tension – all of these have evolved over millennia in one tiny example of how evolution by natural selection 'fits' a creature in the best way to survive in its specific environment.

HOW EYES EVOLVED

Objectors to Darwinian evolution often quote the human eye as an example of something that is so complex that it could not be the result of successive small improvements being inherited over millions of years. I'm not sure why they pick on the eye.

Every aspect of human anatomy and physiology is exquisitely designed to perform a specific task day after day for seventy or eighty years, growing and adapting to circumstances and repairing itself when damaged. The kidney, the liver, the brain, digestion, the blood system – all of these are equally complex and equally difficult to explain, if you know nothing about science, as is the case with many of the critics of evolution.

But there's been a focus on the eye, so to speak, because it's a little easier for the non-scientist to understand how it works and compare it with human artefacts such as cameras, telescopes, microscopes and other optical instruments. We know that these involve sophisticated design and manufacturing processes, carefully machined components including lenses, research into light-gathering materials, servo mechanisms for focusing and so on. How, some people ask, could the biological equivalents of these processes have occurred without any intervention by an intelligent mind with an ultimate purpose in view?

But biologists have a wealth of evidence around them today which makes it quite easy to understand how every stage in the evolution of the eye might have occurred on a timescale that is perfectly feasible, given the duration of the existence of life on Earth. Because, of course, the human eye wasn't created from scratch in the first humans. It owes its current design to an extensive sequence of earlier versions of light-gathering devices in creatures all the way back to the evolution of fish and earlier, over five hundred million years ago.

The earliest stage could have been a random mutation in the skin cells of a creature which made the cells sensitive to light and shade. The offspring of this creature would have a small advantage – if the shadow of a predator fell on the light-sensitive cells, the creatures would be aware and could take avoiding action where the other members of the species would be gobbled up. In the next, improved, generation there would be a slightly larger proportion of the light-sensitive creatures who survived and their offspring would in turn have a greater chance of surviving in the next generation. So far so good. But mutations are happening all the time, and one day, one of these light-sensitive creatures might be born with a mutation that led to the cells being in a shallow pit in the skin. Now there could have been an extra advantage. Not only would the creature know that there was a possible predator near by, it would even know the rough direction of the predator. Instead of just an 'on-off' firing of the light receptors when a shadow passes over them, the messages from different receptors would enable the creature to know the direction of the predator's approach and scuttle off in the opposite direction. As creatures with pits and receptors predominated, any mutation that deepened the pit would increase the accuracy of the rudimentary eye, and lead to a great advantage. Evidence for this type of eye has been seen in fossils and exists today in flatworms and molluscs.

Further improvements could come with the aperture at the top of the pit becoming smaller, creating the effect of a pinhole camera, which could have led to the first stage in actual image formation rather than just light–shade perception.

Some people who are given such an explanation for the first stages of eye evolution, followed by further steps such as lens and retina formation, accept the possibility of this mechanism but cannot understand how several different components can all evolve in sequence to arrive at the right combination of components to work together. 'Half an eye is no good to anyone,' they say, but in fact, as Colin Tudge, the biologist and writer, has written:

Half an eye really is better than no eyes at all. A retina is extremely useful even without a lens to provide fine focus. Even when poorly developed it can distinguish light from dark, and perceive movement. Indeed, a single photoreceptor is useful, let alone a retina. Lenses could have evolved first of all as transparent protectors and developed their focusing power later: at its simplest, after all, this requires nothing more than convexity. Thus as Darwin himself pointed out, we can observe thousands of creatures with far simpler eyes than the human's, down to and including the single eyespots of many protists [microorganisms].

Another concern of people who doubt the power of evolution is the amount of time they believe it would take for the necessary sequence of such tiny changes to take place. In fact, the one thing evolution has in great quantity is time.

Two Swedish scientists, Dan Nilson and Susanne Pelger, carried out a fascinating computer simulation of the effect of random mutations on a computer-generated layer of light-sensitive receptors. With each 'generation' they retained only those offspring that had a slight advantage in light gathering and analysing. By generating random changes in these offspring, and allowing only the one with a slight advantage to go on to the next generation, they could observe what would happen to the primitive eye over successive generations and calculate how many generations would have been necessary to arrive at a spherical eye with a lens and a retina.

Their results were astonishing. With a number of very conservative assumptions, based on what we know about biology and genetics, they concluded that to evolve from flat skin to a functioning eye would take about 400,000 generations. With a typical lifespan of a year or so in the small animals in which eyes developed, this suggests that it would take less than half a million years to evolve a functioning eye. Since complex animals have been around for 500 million years, there has been time for the eye to evolve from scratch many times in any one biological group. And indeed, this is borne out by the fact that biologists have discovered that the eye has evolved at least forty times independently around the animal kingdom.

Far from Darwin's theory of evolution being an improbable means by which complex organs and organisms can arise, scientists would actually be surprised if such complexity had *not* arisen, taking into account mutation rates and the duration of time during which there has been life on Earth.

THEM DRY BONES

Our knowledge of dinosaurs – yours and mine – comes partly from imaginative reconstructions in films and TV documentaries and, to a lesser extent and more accurately, from displays in museums. Often these are accompanied by accounts of dinosaur lives and behaviour which are confident portrayals of creatures that exist today only as bones or casts of bones. How do palaeontologists work out the facts about dinosaur life from a pile of bones?

One puzzle about dinosaurs is their size. Some of them, known collectively as sauropods, were by far the largest animals ever to inhabit the land, with body masses of 50 to 80 tons, ten times heavier than the largest mammals and non-sauropod dinosaurs. And they fed exclusively on plants, dominating their environment for 100 million years, far longer than any other herbivores.

How was it that this particular group of animals, up to 40 metres long and 17 metres tall, could grow to such an enormous size where other animals never evolved beyond perhaps a tenth of the size of the sauropods? And how on Earth could a modern scientist answer such a question faced only with a pile of bones, and very rarely a whole sauropod skeleton?

Recently, two scientists from Germany and Switzerland presented a chain of reasoning which began with a few facts about a sauropod's eating habits and ended with a convincing explanation of dinosaur size.

The starting point was the fact that, from a knowledge of sauropod diet and an analysis of the anatomy of the head and neck, it could be inferred that these dinosaurs did not masticate, or chew, their food, nor did they grind it up in what's called a 'gastric mill', used by birds – descendants of dinosaurs – to swallow food and churn it around with the help of pebbles, which act like millstones. But the vast amounts of plants needed to supply energy to the dinosaurs would have needed long digestion times, and since a large body would include a large gut, this would achieve that aim.

Because the dinosaur didn't need to chew its food, its head could remain small – no need for large jaws and powerful jaw muscles. Creatures with large heavy heads could not evolve long thin necks, but the small-headed sauropods were not limited in this way, and a very long neck evolved which allowed them to forage for food far out of reach of other animals.

Further clues to how the sauropods managed with such large bodies came from discoveries about the animal's respiratory system. Instead of the inspired and exhaled air being confined to specialized sacs, the lungs, dinosaurs, like birds today, had a breathing system called flow-through breathing, which allowed the oxygen in the air to come into contact with the creature's blood in many different areas of the body, including in the long neck. This meant that the air could be used as soon as it entered the creature's head rather than having to travel all the way down the neck to reach lungs. And many of the bones, including vertebrae as large as a metre and a half, were filled with air during respiration, which meant that a creature that breathed in this way would be lighter than a creature of equivalent size that had a mammalian type of breathing.

There are other inferences about the sauropod's life and behaviour that emerge from these few facts. To have a reproductive advantage, the dinosaurs would have had to reach reproductive age very quickly, and that means going from 10 kilograms or so at birth to 100,000 kilograms as an adult, fifty times the newborn size, in twenty years or so. But other animals with high growth rates have a high metabolic rate, the rate at which they burn food and turn it into muscle and other body tissues. If the sauropods had had a high metabolic rate at full size they would have required a huge daily intake of food and could have overheated. It's been calculated that, to avoid this, the sauropod probably had a variable metabolic rate, highest when growing and dropping off as it reached full size.

One final piece of the jigsaw puzzle was filled in by the discoveries of dinosaur eggs, an unusual form of reproduction for large-bodied herbivores. The largest mammals all give birth to single offspring formed in a womb. This made them susceptible to chance extinction if the population density dropped too far and it took a long time to build up the numbers. With egg-laying, sauropods could have many small offspring, so that the chances of extinction were much lower.

Scientists have often looked for external factors to explain the extinction of the large dinosaurs. Meteorites, volcanic eruptions, changes in land-mass size, temperature or carbon dioxide levels – all have been identified as possible causes. But in fact, in the face of many such hazards, the evidence suggests these large creatures actually *resisted* extinction for a very long time. The question 'why did the dinosaurs become extinct?' has been replaced for some scientists by 'how did they survive so long?', and the answer seems to start with the simple fact that they didn't chew their food.

CURIOSITY KILLED THE ELEPHANT

The Etruscan shrew weighs about two grams – three-quarters of an ounce – probably the lightest weight at which a mammalian body can exist. A 4-ton elephant is 2 million times the weight of the shrew and a blue whale is twenty-five times as heavy as the elephant.

When scientists study different mammals, they sometimes have to make comparisons and to extrapolate from the results of research done on one animal to larger or smaller creatures. One attempt in 1962 to apply some results of research on cats and humans to elephants came badly unstuck because of a failure to realize that the size of an animal's body is only one factor in comparisons with smaller or larger animals.

Tusko, a male Indian elephant, was a resident of a zoo in Oklahoma City. Two local psychiatrists were interested in a phenomenon unique to elephants called 'musth'. For a period of days or weeks, the elephant will go on the rampage, becoming extremely aggressive, to such an extent that any human who approaches is in danger of being killed. The musth is accompanied – and may even be caused – by a secretion from glands in the elephant's head which swell and press on its eyes, causing extreme pain. The secretion is said to taste 'unbelievably foul', although this may be an anecdotal report rather than the result of a series of experimental tastings.

Even today, no one really understands the phenomenon, but in 1962 the Oklahoma scientists thought that they could study it by administering LSD, the personality-disrupting drug then in fashionable use. They hoped that it might even lead to the secretion of the unbelievably foul liquid from the temporal glands and confirm the analogy between the madness of musth and the psychosis caused by LSD.

The problem was that the scientists appear to have had a rather shaky understanding of zoology, and the co-author of the paper reporting on the results of the research, although he worked at the zoo, also seems to have overlooked a key step in deciding how much LSD to give the elephant.

If you inject a substance into the bloodstream of an animal or a human, the more blood the body contains the more dilute the substance will be. So since a 2.6-kilogram cat can tolerate a dose of 0.1 milligrams, by working out how many times larger than a cat an elephant is, the scientists calculated that it would be safe to try a dose of nearly three hundred milligrams, three thousand times the safe dose for the cat.

The result was a disaster. The elephant started trumpeting and running around, swayed and fell and died after five minutes.

What had gone wrong? Well, the scientists neglected other factors that are vital in understanding how mammalian bodies handle drugs. Different animals have different metabolic rates, the amount of energy expended while at rest. The higher the metabolic rate, the faster chemicals are broken down in the body. In addition, different animals have different brain sizes in proportion to their bodies, so if you are injecting a drug that affects the brain you need to look at those proportions, too. In fact, once you look in detail at how different bodies handle injected chemicals you can come up with at least five different ways of calculating a safe dose of LSD for an elephant based on studying the effects in different animals. These range from 0.4 milligrams – just twice what was given to the cat – through 3 milligrams, 8 milligrams, 80 milligrams and the largest, and tragic, dose of 297 milligrams.

In fact, even setting aside the ethical problems of this research, which would be much more tightly controlled today, this story illustrates the dangers of overconfidence in science. In concluding their report of this research, the scientists wrote: 'It appears that the elephant is highly sensitive to the effects of LSD.' In fact, a better conclusion would be to say that this particular elephant was highly sensitive to the incompetence of the scientists.

'CAW, I NEVER FORGET A FACE'

We're all aware of how, in spite of the basic similarities between different human faces – same number of eyes, noses and mouths – we can all differentiate and remember many thousands of faces of people we come across. In fact there's even a tiny area of the brain where that ability is located, and if it is destroyed by a stroke or other brain injury people can be in the unfortunate state of never recognizing a face, however recently they've seen it and however well they know the owner.

It's a surprising fact, however, that some animals also have the ability to recognize and remember human faces over a long period. The research that demonstrated this was done on crows. As for all birds, it's important for them to recognize other members of the flock, and so they have a good memory for the faces of other birds. But why should their memory for human faces be any better than mine for the faces of the crows that perch in my trees?

Nevertheless, when a biologist at the University of Washington, John Marzluff, was studying crows he became aware that they seemed to know him and remember him after he'd been away for a while, and he decided to study this ability systematically.

He made two types of mask, designated as 'dangerous' and 'neutral'. He and his colleagues wore the 'dangerous' mask, depicting a caveman, when they trapped and banded the crows or handled them in other ways. But when they were following innocuous pursuits, such as walking on campus, they donned the 'neutral' masks, depicting the then vice-president of the USA, Dick Cheney. (Let's not get into an argument over whether Dick Cheney was actually more dangerous during his term of office than the average caveman.)

By using the masks, the researchers could be sure that the birds responded to their faces rather than to gait and clothing, which would be the same with the neutral mask.

In the following months, although they left the birds alone, the researchers would walk about on campus from time to time wearing either the caveman masks or the Dick Cheney masks. The crows had not forgotten. The cavemen were pestered, scolded and divebombed by the crows, even if the researchers wore their masks upside down. But the Dick Cheneys produced little or no reaction, and they were largely ignored.

The result was dramatic, and Marzluff decided to tighten the experimental conditions to make sure the effect was real. This time he used more realistic masks, denoting one type as dangerous and the other as neutral, and in addition other volunteers wore the masks to test the birds' recognition powers, without knowing which category of mask they were wearing.

Again, the results were unambiguous. One volunteer said: 'The birds were really raucous, screaming persistently, and it was clear they weren't upset about something in general. They were upset with me.'

Researchers in the past had observed similar behaviour among ravens, gulls and other species. The pioneering biologist Konrad Lorenz used to do his research on crows while wearing a devil's costume, so that he could walk around unpestered when he wasn't actually handling the birds. But no one had done systematic research like Dr Marzluff's.

Birds have, and need, sharp visual acuity and the ability to spot and remember detail over long periods. Presumably, their ability to remember patterns, in faces, terrain, the sky or other birds, has become generalized to the distinguishing features of any creature who might be a threat. Their visual acuity is eight times better than humans', even though they have much smaller eyes. But their eyes take up a much bigger proportion of the mass of the head than humans – 15 per cent compared with 2 per cent. A pigeon can pick out a seed 0.3 millimetres long at a distance of 50 centimetres, equivalent to seeing a 60-centimetre post at a distance of a kilometre.

Such recognition abilities raise the possibility that the next time a bird deposits a dropping on your head, the event may be less random than it seems. If it happens often, you could try wearing a Dick Cheney mask. . .

PAPER TIGER, HIDDEN FAKER

When fraud is detected in science, it can often take investigators some time to unravel [🐕 55], since fraudsters who know their subject well can also devise methods that cover up their dishonesty. And yet, when the fraud is uncovered, the method can be disarmingly simple. A famous case in the USA in the 1970s centred on the claim that a black skin graft on a white mouse had not been rejected. It turned out that there was no skin graft, merely a patch of black ink.

In many cases of fraud, the result may be surprising, but it's usually something that scientists want to hear, and therefore are prepared to believe. So initially, the euphoria that greets the discovery can inhibit any sceptical responses.

When Zhou Zhenglong, a Chinese hunter, told of his successful attempts to photograph a South China tiger, which was believed to be extinct, there was celebration among zoologists. The regional authorities had offered a reward for photographic evidence of the tiger's existence. 'It's tremendously exciting news,' said a US conservationist, and the Shaanxi Forestry Bureau pushed ahead with plans for a tiger reserve to make sure the creatures didn't become extinct again.

At a press conference, Zhou described how he had tracked the tiger and taken photographs of it, for which he was paid more than £1,500. 'I guarantee with my head that the pictures are genuine,' said Zhou, and the same Chinese guarantee was used by a forestry department scientist and a member of the Chinese Academy of Sciences.

These days, photographs are all too easy to fake with computer software, but in fact the photos Zhou showed seemed to be genuine and untampered with. They *were* actually taken in the jungle and they *did* show a tiger peering through foliage. The only problem was – as many Internet viewers of the images then pointed out – the tiger in the picture was a paper tiger, taken from a poster or magazine and propped up among the foliage. When botanists looked at the leaves that were at the same distance as the tiger in the photo, they worked out that either the tiger's head was tiny or the leaves, usually a few millimetres wide, had grown to the size of dinner plates. What's more, in several of the photos of the tiger in different poses, the same leaf – which turned out to be part of the photograph – was poised over its forehead, casting the same shadow.

The end of the story was sudden and harsh. In June 2008, Zhou was arrested, seven forestry officials sacked and six more disciplined. Police recovered from Zhou's home a Chinese New Year poster for 2002 with an identical picture of a South China tiger, and a wooden model of a tiger's paw, used to fabricate footprints.

DOGGED BY UNFAIRNESS

Many dog-owners believe their pets are almost human. (Cat-owners *know* this to be true of their pets.) But designing experiments to prove facts about animal perceptions and judgements so that we can compare them with humans is not easy. You can't explain to a dog as you can to (most) psychology students what it is you are researching.

A group of researchers in Vienna were interested in whether dogs had a concept of fairness or equity. It had been shown that some primates display a range of behaviours that suggest some kind of ethical or moral sense. Some primates believe in equal pay for equal work and rebel if a fellow primate receives a better reward for a similar task. Also, a hungry rhesus monkey will not take food if it will lead to another monkey receiving an electric shock. But no one knew if any non-primates had the same sense of fair play.

The researchers obtained forty-three dogs of various breeds and trained them to put their paws in someone's hand on command. The dogs all learned the trick quickly and performed it almost every time they were asked, with or without a reward. Then the researchers took the dogs in pairs, and carried out a series of tests in which they ordered both dogs to do the trick they had been taught, but rewarded only one of them, the same one each time. Pretty soon, it became clear that the unrewarded dog was disgruntled. He required more prompts before he would present his paw, and looked away and scratched himself more often. 'You get the sense that they are not happy,' said the scientist in charge of the experiment. When the results were analysed, it turned out that while the rewarded dog did the trick every time, the unrewarded dog presented his paw only thirteen times out of thirty. Clearly, he did not take kindly to his partner wolfing down a delicious reward each time while he was left on the sidelines. In other words, it was all very unfair.

But did this result *really* mean that dogs have a sense of the inequitable? One further experiment needed to be done before the scientists could be sure of their interpretation of the result. Suppose the unrewarded dog just got bored when he was asked to do the trick over and over again without any reward. This would have nothing to do with any innate awareness that he was being treated unfairly. But in fact, when compared with a solo dog who did the trick and received no reward, the partnered dog still did worse.

Dogs and wolves are well known for collaborating with each other in packs, and successful collaboration depends on everyone pulling his weight. Unfairness in the wild can jeopardize the survival of the whole pack, and so a concern for fair play, rather than every dog for himself, might have evolved to ensure the efficiency of group action.

If dogs really do have a sensitivity to when they are being ignored in favour of other creatures, it might explain what many dog-owners observe when a new baby enters the family and their dog seems particularly put out. We now know that it *seems* that the animal has a sense of unfairness because it probably does.

THE MULTIPLICATION OF SPECIES

The former US defense Secretary Donald Rumsfeld was widely mocked – and why not? – for talking about 'known knowns', 'known unknowns' and 'unknown unknowns', but in fact 'unknown unknowns' have been a useful concept in a number of areas of science over the years. In aeronautics, the phrase has been shortened to 'unk-unks', meaning problems you cannot anticipate because you don't even know they exist.

In answering the question 'How many species of living organisms are there?' unk-unks have played a useful role. This question is usually taken to embrace plants, animals and microorganisms like bacteria. To find out the answer, you could start by making a list of every organism known to the specialists in each field – botanists, zoologists, microbiologists and so on. The total comes to about 1.7 million, a pretty impressive figure when you realize how few of them each of us could name. But of course this total refers only to *known* species. Are we sure we know all of them? Of course we don't. Every day there are stories of scientists exploring new habitats, from rainforests to the depths of the oceans, and finding species they didn't know existed. How can we learn about 'unknown unknowns', and at least turn them into 'known unknowns'?

An entomologist, Terry Erwin, carried out an interesting survey of Panamanian beetles thirty years ago. He chose one tree in Panama and collected every single beetle he could find in that tree. Only a small fraction of all the beetles he had collected had already been identified, and he worked out on that basis, by extrapolating to other organisms, that the true number of all species on Earth was probably nearer 30 million.

Now, Erwin was really only interested in creatures that were visible to the naked eye. (In fact, he was really only interested in beetles, which form the majority of all creatures. This fact led the British scientist J. B. S. Haldane to observe that God seems 'inordinately fond of beetles'.) But if you include microorganisms in the total, there are about forty thousand known species of them. Using similar techniques of extrapolation from the known to the unknown, it had been worked out that there could actually be as many as 400 million different species of microorganism, not including viruses, which are in a class of their own, halfway between living and non-living.

So by using some valid, if a little inaccurate, statistical techniques, we have multiplied the number of species on Earth by two hundred or so.

The writer Colin Tudge has described this process in his masterly guide to all known species, called *The Variety of Life*. He goes on to ask how the number of species on Earth today relates to the total number that there have ever been, and comes up with some mind-blowing figures. Taking as an example a creature more interesting than beetles (unless you're Tom Erwin), he looks at the number of elephant species today – two –

and compares that figure with the number there have been over the last 50 million years. Fossil data show that there have been about 150 different species of elephant in that time. The same exercise for rhinoceroses compares five species today with about two hundred over the last 50 million years. These two calculations suggest that in the past 50 million years there have been a hundred times as many species as there are today. But why stop at 50 million years? There has been life on Earth for at least 3,500 million years. 'It would be surprising,' writes Tudge, 'if the total number of species in the past did not exceed the present inventory by at least 10,000 times,' which gives us a grand total of about 4,000 billion. And that's *species*, not creatures. It's a tribute to the versatility of DNA, the long molecule that governs the design of every creature, that it can ring so many changes and, no doubt, produce another 4,000 billion designs in the next 3,500 million years.

BRAIN AND MIND

CAN MONEY BRING HAPPINESS. . .?

. . .Yes, if you give it away.

Psychologists and other social scientists are often accused of stating the obvious when they report their experimental results. So when a group of psychologists in Canada set out to investigate the influence of income on happiness, they might have been accused of wasting money on a question to which 'most people know' the answer. No one seeks a salary cut in order to have a happier life; when taxes are lowered or the mortgage rate drops, people are generally happy about the changes in their circumstances.

The psychologists narrowed the question down a little. Data from around the world has shown that although real incomes within developed countries have risen dramatically in recent decades, people have not reported a corresponding surge in their happiness levels. It seemed that people were pouring their new money into purchases and pursuits that actually didn't make them any happier. Dr Elizabeth Dunn and her colleagues decided to look more closely at how people spent their money, and see whether different ways of spending money produced different levels of happiness. Their first finding was surprising – people's happiness was correlated with how much money they *gave away* rather than how much they spent.

Now it is a well-known pitfall in science to believe that a *correlation* – two things increasing or decreasing at the same time – means a common *cause*. For example, one scientific study reported that young children who sleep with the light on are much more likely to develop myopia (short-sightedness) in later life. Further investigation showed that short-sighted children were more likely to have short-sighted parents, and short-sighted parents were more likely to leave a light on in their children's bedroom.

So Dr Dunn did another experiment to try to see whether the cause of people's happiness really was due to the fact that they gave money away. About fifty subjects were asked to rate their level of happiness in the morning and then given either $5 or $20, which they were asked to spend by 5 p.m. the same day. Half of the subjects were told to spend the money on themselves and the other half to spend it on somebody else or to give it to charity. After 5 p.m. the subjects were again asked to report on their level of happiness. The results were unambiguous – the people who had been told to spend the money on others finished the day happier than they had been at the start, and the increase in happiness was greater than for those who spent the money on themselves.

In seeking an explanation for why increased income on its own doesn't bring happiness, psychologists have pointed out that it's not the circumstances of our lives – income, gender, religious affiliation – which predict happiness levels, but the choices we make and the practices we choose to engage in. So how we choose to spend our money is at least as important as how much money we have.

One final finding of this research is that giving away quite small amounts, as little as $5, can have a disproportionate effect on the happiness of the giver. The researchers end their report by suggesting that policy interventions at a national level that encourage people to spend more on others rather than themselves may increase the sum total of human happiness.

WHO MOVED MY FINGER?

Most of us would want to believe that, when we decide to perform a simple action like lifting a finger, we make the decision with full awareness in one part of our brain, send an order to the muscles, and the finger moves.

It would be strange – even unbelievable – to have to consider another possibility: that the brain decides before we are aware of it. Somehow, the idea that our decisions occur at an unconscious level and we are made aware of them only after the event, perhaps like someone who is sent – from some unknown decision-maker – a copy of an email ordering the finger to move, is difficult to swallow. It seems to throw into question the whole idea of free will.

Nevertheless, there is fascinating scientific evidence that is difficult to explain away suggesting that our *awareness* of how we make decisions may give us a false impression that we have made the decision ourselves.

The idea that much of our brain/mind activity takes place at an unconscious level is not new. Freud's theories suggest that most of the mental processes that guide our behaviour are unconscious. We might accept that the reason we decide to move a finger is unconscious – to express anger, to achieve pleasure – but we still cling to the idea that the decision is consciously made.

More than twenty years ago, Benjamin Libet, a pioneering US researcher into consciousness, devised the following experiment:

He attached movement detectors to the fingers of volunteers and electrodes to their scalp. In front of the subject was a cathode ray tube with a dot that rotated in a circle. The subject was asked to move his finger at any time, and to note by observing the position of the dot on the screen the point at which he became aware of the decision. During this process, Libet also measured the timing of the subject's brain activity.

The experiment showed the following sequence of events:

1. Half a second before the finger moved, there was brain activity in the part of the brain that sent the instruction to the muscles.
2. A fifth of a second before the finger moved, the subject was aware of making the decision.
3. The finger moved.

Effectively, Libet was observing what the subject was going to decide before the subject himself knew.

These results are consistent and repeatable, but perhaps not as important as they might seem. A difference of three-fifths of a second between making a decision and being aware of it is pretty short and may have explanations that don't breach the idea that we have free will. But supposing the decision was actually made up to *seven* seconds before we became aware of it; surely that really would mean that some 'one' other than ourselves was making the decision and that we didn't have free will?

In 2008, a team of scientists in Germany using sophisticated computer programs to analyse patterns of brain activity reported just this result. They looked at the brains of subjects asked to press a button with the left or the right hand. The choice was up to the subject, who could also choose the moment at which to press the button. The scientists discovered specific patterns of brain activity that occurred up to seven seconds before the subject said he had made the decision about which hand to use. They were so good at recognizing these patterns that they were able to predict which hand a subject would use before the subject knew himself.

The researchers didn't go so far as to suggest that someone else was making the decision. 'Our study shows that decisions are unconsciously prepared much longer ahead than previously thought,' they wrote.

Edward N. Lorenz, the father of chaos theory, would have been unworried by these discoveries. 'We must wholeheartedly believe in free will,' he wrote. 'If free will is a reality, we shall have made the correct choice. If it is not, we shall still not have made an incorrect choice, because we shall not have made any choice at all, not having a free will to do so.'

NOW YOU DON'T SEE IT, NOW YOU DO

In 2004, the Ig Nobel Prize for psychology was awarded to two US psychologists. Their citation says that they were awarded the prize, a trophy and 'little cash but much cachet', for 'demonstrating that when people pay close attention to something, it's all too easy to overlook anything else'.

Scientists are often mocked for the apparent triviality of their researches, for concentrating on tiny details that can't possibly be of use or relevance to society. A US senator, William Proxmire, made a reputation – if you can call it that – by making fun of projects he considered a waste of funding, such as an $84,000 study on why people fall in love, or the SETI (Search for Extraterrestrial Intelligence) project [🐕 79]. One of the projects he mocked, called the Aspen Movie Map, was a pioneering example of interactive computing and has led to many useful applications on the Internet.

The Ig Nobel Awards also seem to pick out projects whose

benefits are not always apparent from a brief description. Their recent recipients include a group of fish experts for showing that herrings apparently communicate by farting; the Vatican for outsourcing prayers to India; two American doctors for their report 'The effect of country music on suicide'; and an old schoolfriend of mine, Professor Michael Turvey, for exploring and explaining the dynamics of hula-hooping. But the stated brief of the Ig Nobels is to mark achievements that 'first make people laugh, and then make them think'.

The psychologists who were awarded an Ig Nobel for their work on attention certainly obtained results that make you think. The problem with me telling you about them is that if I write too much it will destroy the effectiveness of the best demonstration of their discoveries, an effect that at its most dramatic is one of the most extraordinary visual experiences in psychology.

But what I *can* say is that they have shown in a number of experiments a phenomenon called change blindness. A typical demonstration is a clip of video in which a man sits at a desk and a phone rings outside his office. He gets up, leaves the room and the scene cuts to a hallway where the man answers a phone on the wall. Only in the second shot it is a different man, but few people notice this because their attention is 'blind' to this particular change. Other demonstrations include a scene in which over several seconds one particular element is added or subtracted. With such a gradual transition it is very difficult to be aware of the change. Even if the switch between the two versions is instantaneous, which would normally make any changes obvious, when the experimenters put a brief glimpse of a blank screen between the two scenes, it becomes almost impossible to see how they have changed, although once the changes are pointed out it becomes difficult not to see them.

The most vivid demonstration of 'inattentional blindness' is a piece of video on the following website:

http://viscog.beckman.uiuc.edu/grafs/demos/15.html Members of two teams of students, one with white T-shirts, the other with black, pass basketballs between themselves. The task is to count how many times each team carries out a pass. This demonstration has recently been used in the UK in a campaign to promote greater awareness of cyclists among motorists. To find the significance of this video you need to read the note at the end of this chapter. If you read it before you watch the video the effect will be destroyed.

It seems from neurological studies that it is difficult to hold more than four or five separate items in a visual scene in short-term memory. If you are trying to attend to those items, the introduction of other elements will just go unnoticed.

One important implication of this kind of research is that it helps us understand the fact that people sometimes say, after a car accident for example, 'I didn't see it coming', even though the conditions were good, it was clear daylight and nothing was blocking the view. At quite a deep neurological level, their brains would show that they really didn't see it, in the sense of being aware of it.

FACING UP TO BAD CHOICES

A psychology student is taking part in a laboratory experiment to study attractiveness. He's presented with pictures of the faces of two women and asked to say which he finds the more attractive. He chooses face A. The pictures are then put face down on the tabletop and the student takes the one he chose.

He is asked to describe the reasons for his choice, which he does by looking at the picture and saying, 'I chose her because she had dark hair,' or 'I chose her because she smiled.'

There's a problem. In this particular experiment, carried out in a laboratory in Sweden, when the student took the face-down picture to describe his reasons for thinking this face the more attractive one, he was actually given the other picture, face B, by the experimenter's sleight of hand. He then proceeded to give reasons for why he had made his choice, often describing factors that were true for face A but clearly *un*true for face B. The subject who chose a woman because she had dark hair was actually looking at a picture of a blonde woman when he made this statement. The subject who liked smiling women was saying this while looking at a picture of an unsmiling one. One subject even said that he liked women with earrings when shown an earringed woman who *wasn't* his original choice.

The experiment demonstrated what the psychologists called 'choice blindness'. With 120 participants, seventy female and fifty male, the psychologists presented fifteen pairs of photographs, and asked the subjects to choose the more attractive of each pair. But with three of the pairs, the sleight-of-hand switch occurred, so that in describing his reasons for making his choice, the subject was actually justifying the choice he hadn't made.

There were several interesting things about the results. First, only a small proportion, 13 per cent, noticed the switch. The other 87 per cent took it for granted that the face whose attractiveness they 'explained' (with reference to the other face) was the one they had chosen.

Second, the pairs of faces that were used had been chosen so that they spanned the range of similarities. Some pairs were quite similar, others much more dissimilar (as with the blonde versus dark-haired pair mentioned above). And yet there was no difference in the types of explanations offered.

Third, even when they were told of the 'trick' that had been played on them, many of the subjects refused to believe this was the case. This was called by the experimenters 'choice blindness blindness'.

For the scientists involved, these results showed how shaky is the link between intention and outcome. We go through life with certain intentions, we make certain decisions or choices which we think are based on those intentions, and yet the explanations we give for those choices might have little or nothing to do with the real reasons. Another implication of these results is that when we make bad decisions, we rationalize the situation by giving spurious reasons that support the wrong choice.

YOU DO WHAT I DO

In 1983, two American scientists, Andrew Meltzoff and Keith Moore, carried out an experiment to test the ability of people to imitate facial actions. They had forty healthy subjects, eighteen male and twenty-two female, who were required to look at and imitate the faces of the experimenters as they either poked their tongues out or dropped their jaws. This was not an easy experiment to carry out. The scientists started with more than a hundred subjects but, as they say in their report of the results, sixty-seven of them dropped out for the following reasons: 'falling asleep (30%), crying (27%), spitting or choking uncontrollably (24%), hiccuping (15%), and having a bowel movement during the test session (4%)'.

The explanation for this unsocial behaviour is that the subjects were all very young. The youngest was forty-one minutes old and the eldest seventy-two hours, so perhaps the occasional bowel movement was to be expected. The experiment was designed to understand more about a phenomenon that many people had noticed in babies – the unexpected ability, when an adult looms over it making faces, to imitate many of the adult's facial actions. The strangeness lies in the fact that a newborn baby with little or no experience of the world is able to make the conceptual link between a pink piece of flesh moving around in the gap in an adult's face (tongue waggling) and the muscles leading to its own tongue, which it has never seen.

Earlier work on this ability had used older children, and psychologists had suggested that there was some kind of conditioning going on – that a baby would first make a series of random movements and then if, coincidentally, it made the same movement as the adult, this would lead to laughter and smiling on the part of the adult and a quick phone call to Granny. This result would be rewarding to the baby and so the next time an adult poked its tongue out, the baby would be more likely to imitate it. Meltzoff and his colleague thought there was more to it than this. They wondered whether the ability was actually innate, and to find out they had to use babies who had had little or no opportunity for social conditioning, or any other type of behaviour, and as young as possible. Hence snatching babies from the delivery room and neonatal ward – with the parents' permission, of course.

This experiment was the beginning of a long programme of research looking at young children of different ages and the ways they imitate adults. With the care and seriousness of good scientists, they spent hours making funny faces and playing with toys, in carefully controlled situations so that they could produce reliable and unambiguous data. For example, two experimenters would sit side by side opposite a baby, with two video monitors behind the child. One monitor showed an image of the child and what he or she was doing while playing with a toy; the other showed a recording of the previous subject, another child playing with a toy. One experimenter would imitate the current subject while the other imitated the previous subject. The child consistently showed more interest in the experimenter who was imitating him or her compared with the previous child. Without the second experimenter there would be no way of being sure that it was the *imitation* which attracted the child's attention.

The scientists believe that the results of their experiments show that at a fundamental level the ability to see others as 'like me' is present at birth and plays an important part in the development of a child's personality and sense of self. Its usefulness to human society is the ability to feel empathy with others and mentally to 'stand in their shoes'.

The researchers sum up their results as follows:

'They suggest that young infants already register the equivalence between acts of self and other. They do so before the use of language or comparisons of self and other in a mirror. This basic equivalence colors infants' first interpretations of the social world and allows them to imbue the behavior of others with felt meaning.'

At the same time as these behavioural experiments were taking place, scientists in a very different area were discovering a type of cell in the brain which may be responsible for this ability. The discovery was accidental. Scientists studying brain activity in monkeys had found a part of the brain that fired when the monkeys made a certain hand movement to pick up a piece of food. To their surprise, one day they detected this signal from the monkey's brain when it wasn't making the movement but another monkey near by was. Even the act of observing a particular action caused a signal in the same part of the brain that would be active when the monkey was carrying out that action itself. The brain cells that performed this function have been called 'mirror neurons', and the discovery has been described as one of the most important findings in neuroscience in the last decade. Although mirror neurons cannot be observed directly in humans, studies of brain activity using a brain imaging method called fMRI have shown areas of the brain that respond both to activity by the subject and similar activity observed in others. The field is growing, as neurologists and psychologists begin to make links between mirror neurons and language, empathy and even autism.

In archaeology and anthropology, an understanding of the mirror neuron system is being applied to the topic of cultural transmission – the way people learn skills and behaviours from each other. It may be that mirror neurons do not just facilitate imitation but help us infer people's intentions and even their states of mind, a crucial aspect of what sets us apart from animals.

CAN THE BLIND SEE?

Over the last 150 years, medical researchers have come to understand how different parts of the brain perform different tasks in the body. They have located brain centres that are concerned with memory, movement, vision and hearing, touch and smell, speech and so on.

One particular area, in the middle of the back of the brain, is known as the visual cortex because it has been known for a long time that it is where the experience of vision is located. Much of the understanding of how different parts of the brain work has come from the study of injuries. When accidents occur which damage parts of the brain, doctors can infer from the changes in the patient's behaviour or abilities that the injured part of the brain was concerned in some way with the normal functioning that was now disrupted. This is a fairly unsophisticated tool, however. Injuries do not often occur exactly within the borders of different areas of the brain and, in any case, it's a crude source of imprecise information. When a car's lights don't work after a crash, the malfunction could have a whole range of causes, from battery to alternator to lamps or switches, and you have to be very careful not to draw sweeping conclusions.

But with all these reservations, scientists felt fairly sure that when the visual cortex was badly damaged or even entirely removed, the patient would be blind.

But how do you know when someone is blind? This may seem a silly question but some research that started in the 1970s has shown that blindness is not what it seems. The obvious way to test whether someone is blind or not is to ask them. Most people would not lie about such a serious matter, and that is why the link between the visual cortex and the sense of sight was so obvious. No one whose visual cortex had been badly damaged could see, in the sense that we would normally use the word. They were incapable of doing all the things that sighted people do, had great difficulty moving around and, of course, reported a total absence of visual images.

But in 1974 a man known by his initials as D.B. had to have an operation to remove a brain tumour that had invaded his visual cortex, and as a result he was blind in one half of his visual field. This meant that he could see half the world, the half on his left side, but was unaware of anything on his right. Now it so happens that some animals with damaged or missing visual areas can still perform tasks that require the use of vision. They can move their heads in the direction of a flashing light, for example. Since animals can't tell you whether they can see or not, it was not clear what was going on. But faced with a patient who assured the doctors he could not see, a group of scientists led by Professor Larry Weiskrantz at Oxford decided to discover whether D.B. could nevertheless perform similar tasks to those they had demonstrated in animals.

To the researchers' surprise – and indeed the patient's – D.B. showed that there were several things he could do which would not be expected of a blind person. He too could locate a stimulus, a bright light, say, by pointing in its direction. He could also tell whether a stimulus was moving or stationary, and discriminate between a grating that was horizontal and one that was vertical. Most surprising of all, when shown coloured lights in succession he could tell whether their colours were similar or different. And all of this, bear in mind, was done without the patient being at all aware of seeing lights, gratings or colours. In fact, as you might expect, when first faced with these tasks, the patient must have thought the *doctors* had lost their senses. How could someone be expected to see lights and colours in a part of his visual field that he couldn't see? He felt that he was just guessing each time he pointed in a certain direction or answered a series of questions, and was astonished to be told that all his actions and answers proved that, in some sense, he could still 'see'.

What is going on with these patients with this property of 'blindsight', as it was dubbed by Weiskrantz? Well, although the visual cortex seems to be the site of the subjective *experience* of vision, there are other areas of the brain that are involved with the processing of visual messages which are received by the eyes and passed on to the brain. Before a visual message reaches the visual cortex, it passes through other parts of the brain, and even branches out so as to affect several different areas. It has been discovered that as many as nine brain centres are activated by visual messages. What is likely to be happening is that one of those subsidiary centres is capable of directing a hand to point in a certain direction when it receives information about a spot of light, even if the cortex itself has been destroyed. It's like a passenger on a train to a mainline terminus who gets off at an earlier station and hands a package to another passenger on a train coming *from* the terminus. Anyone waiting at the terminus to take part in this handover would be unaware that it had taken place, even though the package would arrive successfully at its destination.

As this phenomenon is studied further, constrained – fortunately – by the rarity of serious brain damage with the necessary specificity, it turns out that other senses localized in other parts of the brain can also give rise to similar phenomena. What has been called 'blind touch' seems to occur, whereby someone with brain damage that destroys all sensation in, say, an arm can nevertheless, when blindfolded, respond successfully to instructions to point to where a particular probe is applying pressure to the arm. There is even a report of 'deaf hearing' in which someone with brain damage in an area that mediates understanding of words, who had lost all ability to understand speech, could nevertheless discriminate between the sounds of familiar words and unfamiliar words.

As with the man with only 5 per cent of his brain tissue [🐕 148], scientists are realizing the tremendous capacity of the brain to perform necessary functions on its own, sometimes without us even being aware of it.

IS YOUR BRAIN REALLY NECESSARY?

In the 1980s a British neurologist, John Lorber, reported on a number of cases of people who appeared to have heads that were largely empty, apart from what is called cerebrospinal fluid, a clear bodily fluid in which the normal brain floats. This fluid is like a shock absorber, protecting the brain from damage if the head is knocked. He made his discovery after a student at the university where he worked was referred to him for having a slightly larger than normal head. Using an early type of brain scan, Lorber found that where a normal scan would show brain tissue filling the entire volume of the head, this man, in Lorber's words, had 'virtually no brain'. This was one of the most dramatic of Lorber's discoveries in an investigation that revealed a number of similar examples of people who possessed merely a thin layer of brain cells just underneath the skull, with the rest of the head being fluid. In effect, they had brains that had as little as 5 per cent of the volume of normal brains.

Of course, in the history of illness and pathology there are many sad stories of people with serious deficits, usually leading to major disabilities. But the astonishing thing about the people Lorber reported on was that many of them were leading perfectly normal lives, in which they raised a family, held good jobs and indeed were unaware that their own heads contained very little brain matter. Some were even chartered accountants.

The people concerned had had hydrocephalus when children. This is a condition in which the circulation of the cerebrospinal fluid around the brain and spinal cord is blocked and so there is a slow build-up of pressure in the brain, which pushes the brain cells outwards against the inside of the skull, like an inflated balloon.

Some of the people to whom this happened are severely disabled, but for reasons that no one can explain, about half of them seemed to have normal intelligence with IQs of 100 (the average in the population) and above. Lorber's work was reported in the 1980s, and some scientists were suspicious of the research. 'Most neurologists don't perform brain scans just because a college student wears a large hat,' said one. But over the years similar cases have been discovered. In 2007, the magazine *Wired* reported 'Brain not necessary for French civil service worker', over a story of a forty-four-year-old Frenchman who went to the doctor because he had mild weakness in one leg, and was discovered to have greatly reduced brain tissue, about 25 per cent of normal, as a result of hydrocephalus when a child. Nevertheless, he was a married man with two children and had a regular job as a civil servant.

The key to understanding how brain tissue can still function under such extreme conditions is the slowness with which the pressure of the cerebrospinal fluid changes the brain's size and structure. No one is suggesting that, at a stroke – so to speak – we can all manage with a quarter or less of our own brain tissue. But somehow the brains of these unusual people have managed to adapt gradually with each small increment of pressure, so that as some parts of the brain were being pushed aside, their functions were taken over by other parts.

Intriguing as these cases are, the answer to the question 'Is your brain really necessary?' is 'yes'. Someone whose brain is only 10 per cent of its original mass still has about ten billion brain cells. While a proportion of our brain cells are necessary to live a normal, reasonably stable life, there are many other brain functions we are not aware of and, perhaps, don't use most of the time, but which come to the fore in emergencies, in rapidly changing circumstances or to access information we may need once every ten years. The evidence that one can maintain a normal family life and hold down a job is evidence only of *some* brain activity, rather than the optimum that a fully normal brain can achieve.

ARE WE REALLY AS CLEVER AS WE THINK?

The pursuit of science is based on the belief that the explanations we seek – for the origins of the universe, the mechanisms of heredity, the nature of gravity or the morphology of protein molecules – are comprehensible to the human brain. But supposing they're not?

We've all come across things we don't understand, however hard we try. Sometimes they are scientific, like almost any description of string theory in physics; sometimes they are literary, like the poetry of Wallace Stevens; sometimes economical, like the theories of David Ricardo. But we usually put our incomprehension down to the limitations of our personal intellectual apparatus or interests. But supposing *no* human brain is well enough equipped to understand the universe. What if we are fooling ourselves by thinking that somehow if humans gather enough data and think about it hard enough we will understand?

While human brains are complex and capable of many amazing things, there is not necessarily any match between the complexity of the universe and the complexity of our brains, any more than a dog's brain is capable of understanding every detail of the world of cats and bones, or the dynamics of stick trajectories when thrown. Dogs get by and so do we, but do we have a right to expect that the harder we puzzle over these things the nearer we will get to the truth?

Recently I stood in front of a 3-metre-high model of the Ptolemaic universe in the Museum of the History of Science in Florence, and I thought about how well that worked as a representation of the motions of the planets until Copernicus and Kepler came along. The model consisted of an interconnected set of giant cogwheels, designed to reproduce the movements of the planets as inferred from their passages across the night sky. The model represents the ideas of Claudius Ptolemy, a second-century AD astronomer, as he tried to reconcile the belief that the movements of the heavenly bodies were based on circles with the fact that their actual movements were not easy to explain in this way.

Nowadays, no element of the theory of giant interlocking cogwheels at work is of any use in understanding the motions of the stars and planets. Instead we have a simple and elegant theory based on ellipses rather than circles, underpinned by Newton's theories of motion and gravitation, which actually explain why the movements have to be the way they are.

The history of science is littered with examples of two types of knowledge advancement. There is imperfect understanding that 'sort of' works and is then modified and replaced by something that works better without destroying the validity of the earlier theory. Newton's theory of gravitation was replaced by Einstein's theories. Then there is imperfect understanding that is replaced by some new idea that owes nothing to older ones. A mythical substance, phlogiston, was invented to account for burning; an all-pervasive medium, ether, was thought necessary to explain how light waves travelled through a vacuum. Later, they were replaced by ideas that provided better explanations, led to predictions and convinced us that they are nearer the truth. Which of these categories really covers today's science? Could we be fooling ourselves by playing around with modern phlogiston? Could string theory be the modern equivalent of Ptolemy's clockwork universe?

And even if we are on the right lines in some areas, how much of what there is to be understood in the universe do we really understand? Fifty per cent? Five per cent? It may be the case that we understand only half a per cent and all the brain power we could muster might take us up to 1 or 2 per cent, in the lifetime of the human race.

Of course, some scientists might ask: why should the capacity of our brains even be an issue so long as we have computers? In 2008, a computer called Roadrunner, built for the US Defense Department – of course – was unveiled as the fastest computer around, capable of carrying out a million billion calculations a second. A British computer scientist estimated that Roadrunner was possibly only five to fifty times less powerful than the human brain. 'Wait another three to five years and it will be there,' he said. And in the future there will be almost no limit to the speed and capacity of computers, as an aid to our brains in carrying out science. But in what sense could computers ever *understand* the universe? In the end, the computer is still only a tool. The only object in the universe, as far as we know, which has understanding is a sentient being, and the sentient beings who are best at it are probably us.

So we're back to the possibility that our pursuit of scientific explanations of the universe may be fruitless. What is certain is that, during that pursuit, we will come across innumerable surprises, many of which – as has always been the case – will lead to new inventions and useful discoveries.

THE FIFTH TASTE

Many of us remember pictures of the human tongue from our school biology lessons with the four taste sensations – sweet, sour, bitter, salty – marked out like territories on a map. Bitter was at the back, sweet at the front and sour and salty arranged down the sides. This turns out not generally to be true. There may be specific concentrations in certain areas but they vary from person to person, and in any case the differences are very slight. It's one of those scientific 'facts' which people could have refuted for themselves over the last hundred years by doing their own simple test.

But more surprising than the non-existent taste map is the fact that up until the end of the twentieth century, people believed that there were only four taste sensations. Nobody – apart perhaps from the Japanese – knew that a *fifth* taste sensation lurked in everyone's taste buds, as basic a taste as sweet or salty but never singled out as separate from the other four. It's as if there was an extra colour in the spectrum, between, say, green and blue, that no one had noticed before.

What is stranger still is that the new taste is so unrecognized by most people that there isn't even a word for it in western languages and it has had to be given a Japanese name – '*umami*', pronounced like 'you-mommy'. The nearest we can get to a descriptive word in English – which helps to remember the name – is 'yumminess', which doesn't sound like a sensory word at all. It seems a bit like having a 'prettiness' visual receptor in the retina. But in fact, a whole range of foods that can be characterized as delicious, potato crisps, for example, have a high level of a chemical called glutamate, which is a principal cause of the *umami* sensation. Many snacks contain a 'yumminess' ingredient, called hydrolysed protein, which we now know acts by stimulating the *umami* receptors.

How could such a receptor have evolved without becoming part of our vocabulary of everyday sensations? Well, one theory is that the *umami* taste is an indicator of foods that are high in protein, essential for survival. So while we don't taste something and say to ourselves 'this is high in protein', we nevertheless want to eat more of it, and the evolutionary aim is achieved.

CAVE ARTIST OR CAVE AUTIST?

The beautiful examples of cave art discovered during the twentieth century in France are believed to have cast new light on the civilization of early man in Europe. First in Lascaux in the Dordogne and then in the Chauvet caves in the Ardeche, lively, colourful, expressive paintings of animals on rock walls led archaeologists and cultural historians to make new inferences about the cultural status of so-called 'primitive' man.

'Each of these painted animals,' wrote one, 'is the embodiment and essence of the animal species. The individual bison, for example, is a spiritual-psychic symbol; he is in a sense the "father of the bison", the idea of the bison, the "bison as such."'

Another expert wrote, 'The first cave paintings . . . are the first irrefutable expressions of a symbolic process that is capable of conveying a rich cultural heritage of images and probably stories from generation to generation.'

A third said, 'This clearly deliberate and planned imagery functions to stress one part of the body, or the animal's activity . . . since it is these that are of interest [to the hunter].'

It's always difficult making inferences about people in the past. The historian Robert Darnton wrote: 'Nothing is easier than to slip into the comfortable assumption that Europeans [of the past] thought and felt just as we do today – allowing for the wigs and wooden shoes.' But the judgements quoted above about the significance of cave paintings are sweeping indeed, and deal with people who lived 30,000 years ago, not a couple of hundred.

A British psychologist, Professor Nicholas Humphrey, has drawn attention to the remarkable similarities between the millennia-old cave paintings and the drawings of an autistic child called Nadia, who was studied by psychologists in England in the 1970s. Humphrey's intention in making this compari- son is to suggest that the similarities between Nadia's drawings and the cave paintings are so close that it may well be that the cave artists themselves were autistic.

Side-by-side comparisons of Nadia's work and cave paint- ings she can never have seen (since the Chauvet works, at least, had not been discovered when she was drawing) are uncanny. A drawing of horses, done when Nadia was three, shows the same impressionistic, overlapping motion as that of a group of horses from the Chauvet cave. Similarly, a drawing of an approaching cow by Nadia is eerily reminiscent of a bison from the walls of Chauvet.

In fact, like that of any good scientist, Humphrey's work is detailed and his conclusions carefully phrased. He is not saying that cave artists *were* autistic but that if a three-year-old autistic child can draw animals with the same verve, expression and realism as in the cave paintings, can a similar mental defect in the cave painters be ruled out? Humphrey's theory is based on a further scientific observation. Like many strongly autistic children, Nadia could not speak at the stage she was producing a succession of beautiful drawings. When she eventually learnt to speak, her precocious artistic ability fell away and she no longer produced such dramatic drawings. Could it be, Humphrey asked, that the community in which the cave art was produced had not yet developed speech or did not use it in the way that so-called 'modern' man uses speech, heavy with symbolism and generalizable to describe any physical or emotional situation? Perhaps, as with Nadia, the development of more sophisticated uses of language somehow wiped out the ability to draw in this impulsive and driven way as words replaced images in the intellectual life of early man.

If Humphrey is right, the 30,000-year-old cave paintings are no less beautiful or dramatic, but they are certainly not evidence of 'a spiritual-psychic symbol' or 'a symbolic process that is capable of conveying a rich cultural heritage of images and probably stories from generation to generation'. They are merely the compulsive scribblings of an observant and impulsive caveman or woman and say nothing about a widespread artistic culture in the society from which the artists came.

WHY DOESN'T THE EARTH MOVE WHEN WE MOVE OUR EYES?

This is another of those questions to which the first response is often 'Why should it?' But if you think about it a little more, when something moves in the external world, its image moves across the retina, the 'screen' at the back of the eye, and that's presumably one way we know that it is moving. If something is stationary in the external world, however, and we move our eyes past it, its image will move across the retina in the same way. Similarly, if the whole external world were to move, in an earthquake for example, its visual impact on the retina would be no different from when the eye moves quickly from side to side. How do we tell the difference?

Perhaps we make some calculation about which is more likely, on the basis of generations of evolution: is it the Earth moving around, which happens rarely, or our head or eyes which are moving, which happens all the time? But then when the Earth *does* move around, even a little bit, in earthquakes, we are left in no doubt about what is really happening, so we can actually tell the difference.

The correct explanation is to be found in a feedback system in the body. You can imagine it as a communication network that sends a message to the brain every time the head or the eyes move, saying, 'Don't worry – this violent motion is not real. It's not the world that's moving, it's just me.'

It's like tests of the emergency alarm system sometimes carried out on US radio stations that are always accompanied by messages which say, 'Do not be alarmed – this is not the real thing.' The day you hear the alarm without the message is the time to start worrying.

But what element of the physiology of the head and eyes is it which sends the message to reassure us that it is we who are moving rather than the world? It could, for example, be that there are movement detectors of some sort in the head or eyes, which are activated when the eye moves. Or perhaps there are position indicators, like the sensors that tell someone

– approximately – where his hand is when he tries to touch his nose with his eyes shut.

In fact, these two can be ruled out by a simple experiment. If you close one eye and jiggle the other up and down with your finger, the world will appear to be moving in synch with your finger movements, which, in turn, are moving your eye. If the message that told you the world wasn't really moving came from movement detectors in your eye, your brain would think the world was motionless, because your eye would still be moving in exactly the same way as if you had willed your muscles to move your eyes. Similarly, since your eye changes position in response to your finger in the same way that it changes position when you move it voluntarily, you would again see the world as still and yourself as moving.

The only difference between the movement of your eyes in those two situations is that when you move your eye with your finger, you are doing it passively; when you move it in the same way using your eye muscles, you are doing it actively. It's the fact that you are *choosing* to move your head or eyes which triggers the signal to the brain that says, 'Don't worry, the world is not moving.' Part of the signal your brain sends to your eye muscles is diverted back to the brain to cancel out the impression of movement. That 'cancel' message is not sent when you jiggle your eye with a finger, because the instruction from the brain to the eye is absent.

BLINDING WITH SCIENTISM

Makers of TV commercials for medicines or detergents often feature pseudo-scientific elements in their ads. Implausibly handsome and clean-cut scientists with clipboards in white coats peer sagely at a rack of test tubes (rarely seen in the high-tech science labs of today). Animated sequences featuring lavatories or digestive tracts include vicious-looking blobs representing germs doing battle with kindly particles of detergent or painkiller. Cosmetics contain ingredients with scientific names like panthenol or miconazole, or even spuriously scientific words like Juvéderm or Celsync. A shampoo that contains 'Pro-vitamin B12 and ginkgo biloba' sounds much more effective at washing your hair than one that doesn't.

Advertising executives believe in the power of 'scientism' – the *appearance* of science (just as 'realism' is the appearance of reality) – in persuading people to buy their products. Now, in an ingenious experiment, two psychologists have proved that people's belief in explanations of scientific topics is enhanced by the use of unfamiliar but impressive scientific images.

A hundred and fifty-six students were given a number of science news articles to read, dealing with research that used a technique called 'functional magnetic resonance imaging' or fMRI. This is a technique that purports to show pictures of brain activity while the subject is carrying out different tasks. The researchers were concerned that such pictures were misleading because when they were used in the media they were portrayed as identifying areas of the brain with quite specific functions such as lying, being in love and believing in God, when in fact the images can only definitely be said to show increased amounts of oxygen in blood flow to areas of the brain.

To see whether people's beliefs in scientific explanations were strengthened by such images, the scientists mocked up some news stories, describing various experimental results from research using fMRI. The students were divided into three groups. One group was given articles that reported on the research using text alone; another was given the same text with the addition of a bar chart or a diagram of the brain, summarizing the data (which were also given in the text); and the third was given the text plus an image of the brain which was said to show illuminated areas that confirmed the data in the article. The stories presented three different conclusions: 'Watching TV Is Related to Math Activity', 'Meditation Enhances Creative Thought' and 'Playing Video Games Benefits Attention'.

The experimental results given in the fake text didn't necessarily justify the conclusions. There were other possible explanations, which the students could have picked up. The scientists were interested in seeing whether identical results were treated as more plausible if they were accompanied by spurious scientific images.

In fact, although the inclusion of bar charts and brain diagrams made no difference to the students' trust of the scientific validity of the conclusions, the brain images did. More subjects agreed with the conclusion of the article when brain images were included than not. The images were viewed as 'proof', and the researchers suggest that there is something special about seeing what appear to be actual photographs of what's going on in someone's head, compared with a diagram.

At the end of their article, having proved that fMRI brain scans deceive people into believing experimental results, the experimenters rather sneakily, I feel, argue that when cognitive scientists write up their experimental results in articles or grant proposals, they should include more brain images, whether or not they are really necessary, to increase the likelihood that their results will be believed, and they will get more funding.

GILBERT WAS RIGHT

Most of us think our political views are formed as a result of thinking about society and how it should be organized, combined with careful consideration of the merits of alternative policies. This in turn is related to what we see as the priorities for a just and cohesive human community.

But when W. S. Gilbert (of Gilbert and Sullivan) wrote that 'Every boy and every girl that's born into the world alive is either a little Liberal or else a little Con-ser-va-tive', he was nearer to the truth, only recently revealed by two groups of scientists.

In 2001 a group of Canadian psychologists discovered that there was a genetic basis for the political opinions of a group of subjects they studied. They took eighty-eight pairs of twins, divided into identical and non-identical pairs. Identical twins have identical genes, since they come from the division of a single fertilized ovum. Non-identical twins come from two separate fertilized eggs, which develop in parallel in the same womb.

The twins were presented with a list of thirty attitudes to different aspects of life and asked whether they agreed or disagreed. These ranged from attitudes to crossword puzzles and chess to views on the preservation of life or the importance of equality in society. The results were clear cut. The identical twins shared far more views than the non-identical twins. What's more, the views that correlated most strongly with genetics were clustered around typical conservative opinions such as attitudes to the death penalty and abortion and to racial discrimination and immigration.

An alternative explanation, that people get their attitudes from their environment and particularly their parents' views, was ruled out by the fact that both sets of twins, identical and non-identical, grew up together in the same type of environment.

In the last couple of years, a second group of scientists, at the University of Nebraska, has carried out research that shows how this genetic influence could occur, by leading to differences in brain physiology that condition how people respond to threats. They assembled a group of forty-six Nebraskans with strong political views, either liberal or conservative, measured on a standard scale. They then carried out a series of physiological tests that measured how the subjects responded to threat. Each participant was shown thirty-three images while wired up to electrodes on the skin that measured the fear response, using the way the skin conducts electricity. Most images were emotionally neutral but among them were three threatening images – a very large spider on the face of a frightened person, a dazed individual with a bloody face, and an open wound with maggots in it. To make sure that any measured responses to these images were to do with negative emotions, the scientists showed a second series of images with three non-threatening stimuli, one of which was characterized in their sober scientific report as a 'bunny'.

The results were unambiguous. Those people who had shown strong support for views that are seen as conservative – such as support for military spending, the death penalty, patriotism, and the Iraq War, and opposition to pacifism, immigration, gun control, foreign aid and so on – showed a greater physiological response to threat. There was a similar difference between the two groups in another physiological measure, the blink response to loud noises, which was more exaggerated in the more 'conservative' subjects.

The source of the physiological response to threats is a small area of the brain called the amygdala – meaning 'almond', from its shape. The activity of this nucleus of cells seems to be determined by our genes, and so therefore is the way in which it deals with organizing the body's response to threats, which in turn relates to how we form our political opinions.

Results like these may suggest why it is actually quite difficult to change people's political views by argument – the amygdala is immune to reason. In countries like the UK and the USA where two major political blocs hold sway and – broadly speaking – political power alternates, the proportion of votes that lead to change is quite small, typically varying from a ratio of 45:55 to 55:45. Very rarely is there a population that contains a predominant proportion of conservatives or liberals, at least when judged by free elections.

From the eye-blinking and skin conductance activities of a group of forty-six Nebraskans, the US scientists come up with a sweeping and thought-provoking conclusion: 'Our research provides one possible explanation for both the lack of malleability in the beliefs of individuals with strong political convictions and for the associated ubiquity of political conflict.'

While W. S. Gilbert may have got there first with this particular insight, his science was not always as accurate. The line following the one about liberals and conservatives is 'When in that house MPs divide if they've a brain and cerebellum too, they've got to leave that brain outside and vote as their leaders tell 'em to'. In fact, the cerebellum has nothing to do with intellectual capacity, but deals with the coordination of motor control in the body. Still, it's undeniably a good rhyme.

NOTE [🐕 139]

Most people concentrating on the task don't see that someone in a gorilla suit walks slowly through the scene, stops in the middle to beat her chest, and continues across the room. Not mentioning the gorilla when describing this scene later is not a failure of memory but of encoding. Only about 10 per cent of people notice the gorilla when they are concentrating on trying to count the passes.

ATOMS AND MOLECULES

THE WORLD'S SMALLEST MUSICAL TRIO

A xylophone, a guitar and a drum each no larger than a red blood cell are the result of research in the burgeoning area of nanotechnology. This technology, forecast twenty years ago by American visionary Eric Drexler and Richard Feynman, theoretical physicist and bongo drummer, has progressed rapidly to the extent that scientists can manipulate atoms and molecules to form shapes, structures and even small machines far below the limits of human eyesight, and not always visible through optical microscopes.

Its possibilities are limitless. Everything that is important to human life is made up of molecules – from genes to microchips; pharmaceuticals to pollutants. Until the emergence of nanotechnology, attempts to reshape or repair atoms and molecules were always indirect, since we couldn't see or manipulate the individual objects. The structures of many of these molecules were known in some detail and scientists could see what changes might be necessary to make a better drug, a more productive crop or a smaller microcircuit, but, metaphorically speaking, their fingers were too large and clumsy to make these changes directly.

Now, techniques are being developed such that scientists can actually manipulate molecules directly and create new structures that have innovative biological functions.

As part of their research into how to interact with these tiny structures, scientists have been looking at using very finely targeted light waves to heat them and thus change their structure. In the case of the 'musical instruments', researchers constructed frameworks that held collections of atoms of silicon of different lengths which would therefore vibrate at different frequencies, like the strings of a guitar. The silicon strings are clamped at both ends and when a beam of light heats them up, the stresses created cause them to vibrate, at a frequency which depends on length. The ultimate aim of this research is to find less energy-consuming ways of controlling electronic circuits, taking over a job currently done by wires which, on a nano scale, dwarf the nano-guitar. One possible use for such devices would be to replace the oscillating quartz crystals that are currently used in mobile phones, and do the same job while using much less power.

There's one small drawback with these tiny instruments, if they were ever to be used to play music. Just as they are too small to be seen by the naked eye – 15,000 of them would span a centimetre – their 'music' is at far too high a pitch to be heard by the naked ear. The musical range is 17,000 octaves above that of a normal guitar.

UNFAIR TO BUIJS-BALLOT?

Poor old C. H. D. Buijs-Ballot missed out on the modicum of fame that has since accrued to Christian Johann Doppler, as a result of an effect which is nowadays a familiar experience for anyone who has been passed at speed by an ambulance or a police car sounding its siren.

It's interesting to realize that until at least the invention of the railway, no one would have experienced the change in sound pitch due to a source of sound approaching and then receding from the listener. I suppose it's possible that someone with perfect pitch might notice the change if he was ever passed by a speeding horseman playing a single note on the trumpet, but such experiences were never common and, in any case, how was he to know the horseman didn't deliberately change pitch as he passed?

The reason Doppler gets the credit for the effect is that he suggested it first, but in connection with light waves rather than sound. He suggested that if stars were approaching the Earth, their light should look bluer than normal, and if they were receding it should look redder. As it turned out, his insight was extremely valuable in later years when the 'spectral shift', as it was called, was used by astronomers to calculate the speed at which stars were moving in the line of sight and, eventually, to confirm the idea that the universe was expanding [🐕 6].

But it was Buijs-Ballot's insight that the same effect might be apparent with sound waves which had much more everyday relevance. Often in science, an unusual observation leads to a search for an explanation. But here, the explanation came first, and Buijs-Ballot needed a way to test whether applying the effect to sound led to a verifiable observation.

The steam railway was in its infancy but it provided the fastest form of travel to test the theory. Over a period of two days in 1845, people living near the railway line between Utrecht and Maarsen in the Netherlands saw a flatbed railcar carrying a band of trumpeters pass back and forth blaring away tunelessly – or at least unmelodically – past a group of musicians with perfect pitch who were making notes about exactly what they heard as the railcar approached and passed them. Their data confirmed that Doppler's theory applied to sound

– the notes dropped from a higher to a lower pitch as the train came past and the magnitude of the drop was related to the speed of the train.

The explanation is simple to understand. The pitch of a note is related directly to the number of sound waves that reach the ear in a second. If the source of the sound is moving towards the listener, more waves arrive in a second than if it is stationary; and if the source is moving away, fewer waves arrive. The same is true, by the way, if the listener is moving towards or away from a stationary source of sound, but this is not such a familiar experience. If you imagine travelling in a boat across a lake with a regular series of waves driven by wind, you will cross more waves in a minute if you are travelling against the wind, and fewer if you travel with the wind, until, if you travel as fast as the waves, they will appear to be stationary.

While Buijs-Ballot missed out on having his name attached to the effect as applied to sound, he did end up with a law of his own, in his main specialism, meteorology. Buijs-Ballot's law states that if you stand with your back to the wind in the northern hemisphere, the low pressure area will be on your left. This is because wind travels anticlockwise in a circle with low pressure at the centre. The reverse is true in the southern hemisphere.

POSITIVE FALLOUT FROM THE BOMB

In the early years of the atomic age, enthusiasm for the new and powerful atom bomb led a number of countries to build their own weapons and, as part of the development process, test them in out-of-the-way places such as Pacific islands. During the 1950s and 1960s, H-bomb tests generated large amounts of radioactivity at test sites, irradiating some local people and test personnel as a result of the still-limited knowledge of the weapons designers. But the radioactivity from the tests had a much more widespread effect as a result of radioactive particles spreading into the atmosphere and being carried by air currents around the globe. People ended up ingesting some of the particles, notably atoms of strontium-90 and caesium-137, which are long-lasting emitters of radioactivity. Even today, long after atomic tests in the atmosphere were abandoned, newborn babies have traces of strontium-90 in their bodies, as measured by scientists who collect milk teeth. Since strontium-90 doesn't exist in nature, these atoms must have come from the nuclear tests.

But another type of atom, carbon-14, which was also produced in atmospheric nuclear tests, has become the focus of a series of research projects which have given a silver lining to the mushroom cloud.

In 1992, an Austrian forensic scientist was faced with a problem. Two elderly sisters were found dead in an apartment in Vienna. Their bodies were mummified and so they must have died some years beforehand, but the neighbours hadn't noticed. Normally, this sad story, which seemed to involve no crime, would not have held the attention of the police or forensic scientists. But the women each had pension funds and insurance policies with different companies, and whichever sister survived the other – since it didn't seem that they died at the same instant – would have inherited her sister's pension and life insurance and the bulk of those funds would have gone to her own insurance company.

Archaeologists use carbon-14 to date archaeological finds but this technique is accurate only to within a few hundred years. But two physicists at the University of Vienna had another idea. Levels of carbon-14 in the atmosphere vary from year to year. Between 1950 and the late 1960s, owing to the atomic tests, they rose to a peak and then dropped off more slowly to a level that is still higher than at any time before 1950.

Carbon-14 in the atmosphere is taken into body cells, and the levels indicate the amount of carbon-14 that was in the atmosphere at the time the cells were formed. Some cells in the body stay the same through life and so their levels will indicate the date of birth. But other cells divide and redivide so the levels of carbon-14 in those cells will match the year in which the cells were created.

In the case of the Viennese sisters, physicists tested fat cells in the sisters' bones, which would have been generated just before the sisters' deaths, and they were able to tell that one sister had died in 1988 and the other in 1989, and settle the issue of the inheritances.

It's an interesting story without apparently earth-shattering consequences. But twenty years later, the use of biological levels of carbon-14 from atomic tests as a dating technique has blossomed. One group of scientists discovered that when DNA divides, it effectively takes a snapshot of the level of carbon-14 at the time, and they were able to use this new knowledge to show that in certain regions of the brain no new brain cells are formed after birth, an idea that many scientists believed but had not been able to prove. Carbon-14 levels in teeth, another body feature that doesn't generate new cells, can be used to determine the date of birth.

And at the University of Adelaide in Australia, scientists with a professional interest in wine have used the new technique as a foolproof means of identifying the vintage of a wine by analysing the alcohol's carbon-14, which is derived from sugar made in grapes that took up carbon-14 from the atmosphere in the year of growth.

AN UNCERTAIN FUTURE

Many people have heard of Heisenberg's Uncertainty Principle. This states that in observing elementary particles it is impossible to determine both the position of a particle and its velocity. In order to observe at all you have to use some kind of procedure that affects what you are looking at. If you imagine a world in which we could perceive objects only by touch, the mere action of poking at an object to see where it was could change its position. If it was moving, the 'poke' would also change its speed or direction of travel.

It's a bit like this with atomic particles. If we try to shine a beam of light or other energy on the particle in order to 'see' it, the moment the beam hits the particle it changes one of its characteristics – its position or velocity – from what it would have been if we hadn't looked.

This is very much an oversimplification. It sounds as if it is somehow the measuring equipment which is at fault, when in fact, for Heisenberg, the Uncertainty Principle reflects a basic fact about atomic particles, that physical states with definite position *and* momentum don't exist. Once you measure one the other becomes indeterminate.

While this was an important observation for particle physics it's always presented as having limited relevance to the world of big objects, such as tables and cricket balls and dogs and so on. These are made up of so many billions of atoms and the uncertainties in measuring any one of them are so small that with the velocities and distances that are of interest to us in the macro world, any effect of the Uncertainty Principle was thought to be undetectable.

Before anything was known about the physics of atoms, the French mathematician and astronomer the Marquis de Laplace envisaged a 'demon' with a complete knowledge of the precise position and velocity of every atom in the universe and said that, with sufficient calculating power and using Newton's laws, he could calculate the entire course of the future of the universe. Even after Heisenberg's insight, it is often believed that with a simple physical system, such as a billiard table and balls, it is possible to calculate the future path of the balls as far ahead as you like, equipped with an accurate knowledge of their starting positions and velocities and using Newton's laws of motion.

In fact, this is not true. An American physicist, D. J. Raymond, calculated forty years ago that you could predict reasonably accurately the paths and velocities of the balls on a billiard table for the first *eleven* collisions, and after that, however precisely you did your measurement, the combined uncertainties associated with each of the collisions would be magnified with each collision until it would be impossible to say what would happen after the eleventh.

To give an example of his reasoning: however accurately we aim the cue ball at the first ball we want to strike, we cannot measure the position of that ball with an accuracy below a certain very small figure, so there will be a tiny error in our estimate of where the ball will go after we hit it. This very small mistake will produce a disproportionate error in how the next ball moves and so on, and so this uncertainty, which started as a minuscule figure, will be magnified dramatically with further collisions. Raymond made certain assumptions about the mass and radius of the balls, the distance between collisions, and so on, and worked out how many collisions there would have to be before the errors built up so much that it would be impossible to say with any certainty where any of the balls were. That's how he arrived at the calculation that after eleven or so collisions it would be impossible to forecast the motion of the balls with any useful degree of accuracy.

Of course, in the real world, there are so many larger sources of inaccuracy than these quantum uncertainties – irregularities in the billiard table or the balls, the shakiness of the player's hand, even the effects of small changes in temperature – that realistic prediction will fail much earlier than the eleventh collision. But Laplace would be astonished to learn that not only could his 'demon' not predict the future state of the universe from a comprehensive knowledge of its current state, he wouldn't even be able to predict the behaviour of a few billiard balls over the following ten seconds or so.

WHY SHOULD WE FALL THROUGH THE FLOOR?

Scientists have the habit of asking questions that seem stupid to the non-scientist, and one of them is 'Why don't we fall through the floor?' There is even a book of this name. But in our everyday experience, floors are solid surfaces whose only function in life is to stop us falling through them. It is the least we can ask of them. What underlies such 'stupid' questions, however, is an understanding of floors and how they are constituted which conflicts with our commonsense understanding of such matters.

In this case, since floors are made of atoms it is an understanding of the nature of atoms which has led to this question.

The concept of atoms as basic constituents of matter has been around for 2,500 years, but for most of that time atoms have been imagined rather than understood, as solid hard balls. It was only in the twentieth century that a more nuanced picture of the atom emerged, a picture that is full of puzzles and shows the limitations of trying to think of physical and mathematical concepts as pictures at all.

Nowadays, the 'picture' of an atom based on a century of scientific research is of a hard nucleus one unit across (we'll come to what I mean by unit in a moment) surrounded by a cloud of much lighter particles, electrons, out to a radius of about 30,000 units or more. More than 99.9 per cent of the mass of the atom is concentrated in the nucleus and so between the nucleus and the outer 'surface' there is very little else. Even the electrons are not really 'there'. Although they can sometimes behave like tiny solid balls themselves, they are actually understood by scientists as a cloud of varying density with the densest part of the cloud being where the electron is most likely to be found. The only time an electron 'exists' is if a scientist using some delicate piece of apparatus is able to bounce another electron or some radiation off it. At that point, not only would its position be known, it would actually materialize as a result of the observation, before disappearing again into a cloud of probabilities.

It's as if an individual MP living in England were represented by a cloud that was very dense in the area of her home and the House of Commons, less dense in the shopping mall near her home, even less dense around her gym, and extremely rarefied in the slums of Glasgow or at the top of Ben Nevis. But she herself would be nowhere, until someone actually bumped into her, when the cloud would condense instantaneously to reveal a forty-year-old woman wearing a smart suit.

So back to the atom. If the nucleus were the size of an orange 10 centimetres across, the edge of the electron cloud – the point beyond which the probability of finding an electron was vanishingly small – would be between 3 and 30 kilometres away. So in any solid made up of atoms – a floorboard, say – 99 per cent of the volume is empty space, like a vast plain with oranges scattered several kilometres apart, and electron clouds with very little mass drifting around.

In reality, setting aside oranges and kilometres, the unit that is used to measure atoms is called a femtometre, one thousand million millionth of a metre. The nucleus is about two femtometres across, and the atomic radius is about 100,000 femtometres, so it would take 10 billion atoms to form a line a metre long.

Now, at least, we might be able to see the point of the question 'Why don't we fall through the floor?' In fact, it may have more point than you might at first think. Not only is the 'solid' floor largely empty space, the shoe or foot that stands on it is also made up of atoms, with the same vast proportion of emptiness to mass. Why don't these two assemblages largely made up of empty space just pass straight through each other, like two fleets of ships spaced several kilometres apart proceeding in opposite directions down the English Channel?

The answer to the question lies in the commonsense – but mistaken – definition of what 'solid' means. If we take the word to mean a space entirely filled with mass, then floors can seem very insubstantial. But we need to build into our understanding another characteristic of atoms, force fields that surround the nucleus. When two atoms are near each other they are affected by two forces, one that attracts and one that repels. As one atom approaches the other, it reaches a point where the two forces exactly balance, and it is held firmly in that position. In an assembly of atoms, locked together by these forces in a solid like a floorboard, any other atom that tried to prise them apart would also reach a point of equilibrium where to go any further would mean overcoming a strong repulsive force, far greater than the gravitational force pulling the foot down to the floor.

So next time you walk across the room, imagine that your feet and the floor are pushed a tiny distance apart by the strong repulsive forces of the atoms, and *that's* why you don't fall through the floor.

CRITICAL MASS FOR LOUIS SLOTIN

Successful scientists are often adventurous in their thinking, which has few negative consequences beyond the possibility of seeming silly or outrageous. But adventurous experimentation can lead to far worse consequences, as one young Canadian physicist, Louis Slotin, found out when he was working at Los Alamos on the atom bomb project during the 1940s. In the course of his work he became a victim of the process he and other scientists were trying to understand.

One of the important physical facts that needed to be discovered in order to make a successful atom bomb was a figure called the 'critical mass' of enriched uranium or plutonium, the substance that under certain circumstances would experience an explosive chain reaction [🐕 38]. While this figure could be calculated theoretically, it was essential to have experimental measurements as a check on the calculations. Slotin carried out a number of tests to establish this figure. He would take two pieces of plutonium the size and shape of a split cricket ball and bring them slowly together.

In a chain reaction, atomic particles called neutrons are emitted from radioactive elements like uranium or plutonium. Normally, with a small amount of the element, many of the neutrons escape, but a few of them collide with other atoms and cause the release of more neutrons. Again, most of these neutrons will escape but a few will cause yet more to be released. Below a certain mass of plutonium, the release of successive waves of neutrons just causes heat and low-level radioactivity, and the rate at which neutrons are released dies away. But if the mass of uranium or plutonium is large enough, the rate at which neutrons will cause other neutrons to emerge from atoms will increase and became self-sustaining. The number of neutrons colliding with atoms will go up and up until the whole mass of material is consumed in a nuclear explosion.

The atom scientists called the process of determining the critical mass of radioactive elements 'tickling the dragon's tail', and Slotin was an experienced tickler. He performed more than fifty experiments in which he slowly moved together two subcritical masses, observing the increase in the release of neutrons by listening to the clicks in a neutron counter. As he moved the masses towards each other and the clicks became more and more frequent he had to judge the right moment to stop the movement just before a chain reaction occurred and measure the gap between the two masses, which would allow the physicists to calculate the total mass of plutonium that would lead to a runaway chain reaction.

On 21 May 1946, Slotin was carrying out this test for the last time with a particular lump of plutonium, before it was scheduled to be detonated. There were several colleagues with him as he manoeuvred an outer cover for the two half-spheres of plutonium, designed to reflect neutrons back into the chain reaction to increase its speed. In doing this, he kept the two half-spheres apart with a screwdriver in one hand while wrestling with the cover with the other. (He had removed two shims that had been keeping the plutonium apart.) Suddenly, the screwdriver slipped and he dropped the cover. The two pieces of plutonium snapped together, causing a critical reaction and a burst of lethal radiation. The scientists in the room saw a blue glow and Slotin, the nearest to the reaction, experienced a sour taste in his mouth and an intense burning sensation in his hand. In that instant, he knew that he was doomed. He had received a dose of radiation that was equivalent to standing 1,500 metres from the detonation of an atom bomb. He also knew that his colleagues had received large doses too. Making a quick sketch map of the positions of everyone in relation to the site of the critical reaction, he hustled his colleagues into two jeeps and they all headed for the hospital.

Nine days later, Slotin died, after futile attempts to treat him with blood transfusions from many of his – non-irradiated – colleagues. The other people in the room at the time suffered some after-effects but none of them died.

The incident was classified as top secret at the time, but in later years, when the story came out, the picture painted by Slotin's friends was of a hero who gave his life to save his colleagues. An article about the accident, published in 1995, was headed 'A young Canadian scientist gave his life to save his friends when an experiment went wrong'. An account written by a friend of Slotin's in 1956 says: 'Almost as if by reflex action Slotin hurled himself forward and tore the reacting mass apart with his bare hands. The others gasped and, turning around, Slotin, his face whitely reflecting his terror, motioned them to leave the room. . . The young scientist gave his life, just as did many of his comrades in arms.'

In fact, it came out later that when Slotin's colleagues suggested that he should use some kind of safety device to push the hemispheres apart if an accident occurred Slotin said: 'If I have to depend upon safety devices I am sure to have an accident.' In the years after his death, it became apparent that Slotin had been a fantasist, inventing stories of a dramatic personal past that had never existed, and an inveterate risk-taker. On one occasion, a former colleague said in 1993, Slotin had asked for a reactor to be shut down so that he could make adjustments to an experiment at the bottom of a tank of water, used to absorb radiation. When this request was denied, he came in at the weekend, stripped down to his underpants and dived down to the bottom of the tank while the reactor was still active.

It seems that Slotin's accident was actually an unnecessary death for a foolhardy scientist.

TALL STORY

If you've ever watched a skyscraper or any tall building being constructed, you'll know that an early stage in the process, after the foundations have been laid, is for a structural skeleton to be assembled on which floors, walls and ceilings will be hung. If it's a steel skeleton, as opposed to reinforced concrete, the construction team will start at the foundations by fixing a framework of steel columns several storeys high, linked by beams. The columns at the bottom of the building are the thickest and the heaviest because they will end up having to support the entire weight of the building. As the lowest framework is completed, more columns are added to take the skeleton to a higher level. The columns here are not as massive because they have to support only the floors above. And so it goes on, to the top of the building, with columns getting narrower and lighter as they go up.

As buildings get taller and taller – a kilometre-high building is planned for Dubai – the columns at the foot get heavier and heavier because they have more weight to support. But with the tallest of the new buildings, such as Freedom Tower, the skyscraper being built on the site of the World Trade Center in New York, an extra and surprising feature creeps in that engineers have to deal with when they design these behemoths. In addition to having to stand up to the pressure of a huge weight pressing from above, the columns have to be designed to withstand *upward* forces in the opposite direction to the pull of gravity. There are going to be occasions in the life of these buildings where the columns at the foot are supporting no weight at all or are even trying to hold the building down.

How can this come about? Well, the taller the building the more it is likely to sway in the wind. When strong winds come up against a building in their path they divide and go to left or right. When they do this, they create vortices, like whirlpools, which push and pull the building in a pattern than can make the building begin to sway like a slowed-down tuning fork. In fact, it is better if a building *does* sway – this allows it to absorb some of the wind forces. But when it leans to one side, this will tend to exert a force that counteracts gravity on the columns on the other side of the building, and if the wind force is strong enough it will pull the column upwards, rather than just lessening the weight it has to carry. What's more, in another new factor for engineers to think about, the column on the side the building is leaning towards will have to be designed to take a *greater* weight than the weight of the building, because – as with a seesaw – when one side goes up the other must go down.

And the direction of sway can be surprising. You might think that if the wind pushes against the north face of a building, say, it would move to the south if the wind were strong enough. In fact, the sway caused by vortices would move the building from side to side, at right angles to the wind, and if the wind and the vortices it caused were strong enough to push the building over, it would eventually fall to east or west.

THE RICHES OF YTTERBY

The chemical elements are the building blocks of all matter. By the beginning of the eighteenth century about fifteen had been discovered, some of them well known because they were metals, such as iron or copper, or substances used in primitive manufacturing processes, such as sulphur or phosphorus. From 1700 onwards, as the science of chemistry became more sophisticated – and more rational, replacing alchemy – scores more elements were discovered. By their nature, chemical elements are widely distributed around the Earth, in the ground, the air and the sea. But where they were first discovered was the result of pure chance. In 1825–26, a French chemist, A. J. Balard, discovered the element bromine in seawater, at his laboratory in Montpellier. Round about the same time, C. Löwig in Germany discovered the same element in mineral salts.

But one small spot on the Earth's surface has given birth to seven chemical elements. In 1794, in a quarry on the Swedish island of Resarö, a Finnish scientist, Johan Gadolin, discovered a chemical element that he named yttrium, after Ytterby, the village near to the quarry. Over the next hundred years, six more elements were discovered in the same quarry. With a certain lack of imagination and with some potential for confusion, the next three of those elements were also named after the village –terbium, erbium, and ytterbium. As yet more new elements emerged from the same ground in the late nineteenth century, the namers struck out and gave us gadolinium, after Johan Gadolin, holmium, after the Latin name for Stockholm, and finally thulium, the old word for the Scandinavian countries.

WHAT IS SMASHED IN AN ATOM-SMASHER?

The popular image of an atom-smasher – or particle accelerator – is of a device that somehow smashes atoms together with such high energy that they break apart and we can see what's inside them. There are aspects of 'atom-smashing' experiments which can give that impression. Atoms and some of the particles that make them up, such as electrons and nuclei [🐕 176], are certainly sent hurtling towards each other at extremely high speeds; they certainly collide; and there certainly emerges from that collision a range of particles that were not there before and could be thought to come from 'inside' the atom.

You could think of it as two double-decker buses that hurtle towards each other at high speed and when they collide they break up into seats and engines and wheels and windows and, I suppose, passengers and drivers. But in fact, it's not like that at all. Physicists are building particle accelerators that cause collisions at higher and higher energies in order to *create* particles that have never been seen before. The results of such collisions can be as surprising as if two double-decker buses collided and the result was a white Ford Mondeo, a black Harley Davidson motorbike, an electric cooker, three wooden dining tables, a hundred teacups, six Harrods hampers and a ton of chalk.

What is happening, if the experiment goes well, is that the collisions occur at such speed that all of the mass of the particles is turned into energy. One of the most significant of Einstein's insights was the fact that mass and energy are equivalent, and can be changed from one to the other. His equation $e = mc^2$ is the exchange rate for that equivalence, just as you might say $£ = \$c^2$, with c being the number 1.25, so that c^2 is 1.56, showing that you can exchange a pound for 1.56 dollars. If you change £100 into dollars, you can then change the $156 into any other combination of currencies you like and it is irrelevant that you started with pounds.

So the first result of the collision of particles is the disappearance of the particles altogether, just as your pounds disappeared. They are replaced by a burst of energy – a bunch of dollars. Then this result is followed immediately by that energy being converted into particles. Since the burst of energy has no 'memory' of the particles in the original collision, it is free to turn back into whatever particles it likes, consistent with the 'exchange rate' between energy and mass. If the experiment is done often enough, among those particles produced as a result of the high energy collisions will be new types of particles that don't exist at all in our universe today. Using this knowledge, physicists are employing particle accelerators to recreate the conditions that existed in the early moments of the universe

[🐕 7] when they believe a whole range of other particles were created and destroyed in the high-energy environment.

SEEING NEUTRINOS

When you consider how much money and ingenuity are spent on building particle accelerators [🐕 184] with the sole purpose of detecting some of the tiniest particles of matter, it's surprising that one of the smallest and lightest of particles, the neutrino, can be observed by anyone with a pair of eyes. In fact, one eye will do.

Neutrinos are strange particles, more mysterious than the run-of-the-mill constituents of matter such as protons, neutrons and electrons. Those three types of particle have a mass. Although the electron is 2,000 times lighter than a proton or a neutron, it's still possible to measure its mass. But neutrinos are so much lighter that for years nobody was able to detect them at all. Scientists believed they had to exist because they were necessary to make some physics equations balance out, but for years there was no known way to demonstrate their existence. The fact that they were so light was part of the problem. To all intents and purposes they had no mass at all. When you consider how tiny the masses of 'normal' atomic particles are, the neutrino weighed in at less than a ten thousandth of the mass of the electron.

This had the following result: if a neutrino comes across normal matter it goes right through it and out the other side. This may seem surprising. It seems the opposite of what you'd expect. If you imagine a bullet and a peanut fired from a gun at a block of concrete you'd expect the heavier item, the bullet, to go through, while the peanut was stopped dead. The clue to the neutrino's behaviour lies in the fact that so-called 'solid' matter is actually largely empty [🐕 177]. But some of the components in that 'empty' atom have electric charges, and so if you fire, say, a proton, a charged particle, into a collection of atoms, sooner or later it will travel close enough to be deflected and may be absorbed by the electric charges from within the atoms. A neutrino, as well has having virtually no mass, also has an electric charge perhaps a thousand times weaker than the electromagnetic forces that affect the more massive and more highly charged atomic particles. So defective is a neutrino in the mass and charge department that it could travel through a light-year-thick layer of lead and come out the other side unscathed.

While the chances of any one neutrino interacting with any other particles are very small, there are many neutrinos arriving at the Earth's surface all the time from the sun. Every square centimetre on Earth facing the sun receives 70 billion neutrinos a second. The vast majority of these pass straight through the Earth and out the other side, but occasionally, if the circumstances are right, an interaction occurs in matter and an electron travelling faster than the speed of light emerges as a result.

You may feel that it is time to take stock. What business does something have – even if it is as small as an electron – travelling faster than the speed of light? If you know some physics you'll know that *nothing* can travel faster than the speed of light – in a vacuum. But the speed of light in some other media, water, for example, is normally slower than the speed in a vacuum, perhaps as slow as half the maximum speed, and it's possible for particles to exceed that slower speed. If they do so, they cause a burst of blue radiation, almost in the way a sonic boom is created when something travels faster than sound. To detect neutrinos, scientists use a very large container of some dense transparent substance, such as heavy water or, sometimes, dry-cleaning fluid. Since every part of the Earth is bathed in neutrinos all the time, if physicists wait long enough for the right combination of circumstances they sometimes detect the telltale burst of blue radiation that can only be due to a neutrino interacting with the transparent substance.

So what about the idea that the human eye might be able to detect a neutrino? Well, it's all a matter of probability. The eye acts like a tank of a transparent substance which could experience the same burst of radiation if just one of the 70 billion neutrinos passing through it in a second were to trigger the effect. Occasionally, the Earth is bombarded with neutrinos from farther away than the sun, from an exploding star known as a supernova [🐕 21]. As a result of such an explosion in 1987, it's been estimated that between one and five thousand people would have experienced a neutrino 'event', and if the site of that event was the eye, a handful of them would have seen the blue flash that marked the passage of the neutrino. It's likely, therefore, that the everyday neutrino bombardment of the Earth will cause a similar blue flash in the eye of someone, somewhere, at some time. But the chance of anyone noticing it, reporting it and claiming a neutrino sighting is as slim as someone would be who was fed entirely on a diet of neutrinos.

CLOCK INSIDE THE ROCK

'This rock is 4.4 billion years old,' said Simon Wilde and his geologist colleagues in 2001 'Happy birthday.' (Actually I made the last bit up.) He was talking about the age of the earliest piece of the Earth, but how did he know?

The most accurate way to date rocks depends on zircon, December's birthstone, a semi-precious stone that can be found in a range of colours, as well as in a clear form that resembles diamond. Its physical properties have led in recent years to it becoming one of the best ways of dating rocks accurately. Zircon is everywhere, as very small grains in most types of rock.

For a period beginning about 4.5 billion years ago when it was formed, the Earth suffered an intense bombardment by meteorites, which generated a huge amount of heat, melting the materials that made up its surface. When molten rock cooled, some collections of atoms, including zircon, formed crystals which have survived unchanged until today. Unchanged, that is, apart from one useful characteristic. The lattice of a zircon crystal occasionally accepts a uranium atom instead of another atom of zircon, so that scattered through a crystal of zircon at the moment of its formation are a certain number of uranium atoms. But uranium atoms decay at a known rate into lead atoms. What's more, when zircon crystals form they contain no lead atoms. From the moment of its crystallization, a zircon crystal will steadily acquire an increasing proportion of lead atoms as the uranium atoms decay.

So a zircon crystal in newly formed rock is like a clock set to zero. The 'ticks' in the clock are the uranium atoms contained in the crystal as they change to lead atoms. These changes don't happen at regular intervals but they do happen in a statistically predictable way. Uranium has what is called a half-life [🐕 38], which means the time it takes for *half* the uranium atoms in a sample to change into lead.

One type of uranium, called uranium-235, has a half-life of

704 million years, while another type, uranium-238, has a half-life of 4.7 billion years.

So if a grain of zircon starts with 100 U_{235} atoms, say, and then one by one each atom decays to become a lead atom, after 704 million years the ratio of lead to uranium will be equal and there will be fifty of each. A scientist measuring this particular grain of zircon will be able to say that it is 704 million years since it was formed. After another 704 million years, half of the fifty uranium atoms will have become lead and the ratio will be 75 per cent lead to 25 per cent uranium and the scientist will know that the rock was formed just over a billion years ago. Using the two types of uranium atoms and their ratio to lead atoms, geologists can work out how long ago the process started and therefore how old the crystal is.

In 2000 in Western Australia, at a site called Jack Hills, a deep purple crystal of zircon was discovered, less than a quarter of a millimetre across. It was given the prosaic name W74/2-36 and the uranium/lead ratio showed it to be 90 million years older than any previously dated rock on Earth, at 4,404 million years old. Although the crystal was so tiny, a wide range of other investigations could be made, including oxygen isotope measurements and rare Earth analyses, measurements that provided information about the physical processes that led to the formation of these very old rocks in the area, and even revealed that water was involved in the way the Earth's crust was formed and that there were therefore oceans as well as land at that very early stage in the Earth's history.

SICKNESS AND HEALTH

TICKING BOXES SAVES LIVES

Newspaper stories about medical advances often use the word 'breakthrough'. If the 'breakthrough' allows the journalist to write of 'miracle drugs', 'million-pound scanners' or 'life-saving surgery', all the better. Without such glamour, significant advances can go unnoticed by the wider world. In 2001, a new idea came along that used no new drugs, no new pieces of expensive equipment and no new surgical technique, and yet produced significant changes in patient health and survival in a very short time.

Dr Peter Pronovost at Johns Hopkins Hospital in Baltimore had an idea that, in one hospital in the space of a year, saved twenty-one lives and millions of dollars. It also got him named as one of *Time* magazine's '100 Most Influential People for 2008'. His idea started from the fact that patients get ill and sometimes die if they acquire an infection through the line used to carry drugs and fluids into their blood vessels. So, he thought, if doctors and nurses were to wash their hands, clean the patient's skin with antiseptic and use sterile masks, clothes and dressings, perhaps there would be fewer infections.

A hundred and forty years after Joseph Lister showed the immense value of antiseptis precautions for patients in hospital, Dr Pronovost's observation might seem a little behind the times. But the key element in his work wasn't so much the knowledge that infection travels into the body through unclean procedures as the observation that doctors and nurses regularly *ignored* simple procedures like hand-washing, sterile garments and so on. This was pretty surprising, but it was confirmed by a survey carried out over a month, when nurses observed that a third of doctors failed to carry out one of these simple procedures.

Johns Hopkins is one of the leading teaching and research hospitals in the United States, so it can easily be imagined that surveys in other hospitals might reveal an even higher rate of slippage in basic life-protecting practices.

Having observed the importance of measures that every doctor was meant to know about, and seen how far short his colleagues fell of what was required, what was Dr Pronovost's big idea? He devised a checklist to be used every single time a doctor put a line into a patient's body. Each of five items had to be carried out and ticked on a list. Doctors had to:

1. wash their hands with soap;
2. clean the patient's skin with chlorhexidine antiseptic;
3. put sterile drapes over the entire patient;
4. wear a sterile mask, hat, gown and gloves; and
5. put a sterile dressing over the catheter site once the line was in.

When Pronovost published his checklist in an American medical journal, it was adopted by intensive care units in hospitals in the state of Michigan. Within three months the state's infection rates in ICUs had dropped to a third of their previous level, and by the end of the year many hospitals had ICU infection rates of zero, outperforming 90 per cent of the nation's ICUs. Within eighteen months of starting to use Pronovost's checklist, deaths from infection had dropped by 1,500 and it was estimated that the hospitals had saved $175 million in costs of care.

Remember, these dramatic results had been obtained through using one simple five-item checklist in one type of medical procedure. Pronovost and his colleagues are now devising other checklists for use in surgery, stroke treatment, cardiac care – in fact, every medical intervention where a lapse of memory or a moment of distraction could lead a doctor or a nurse to omit one key step in the time-consuming and sometimes irritating list of tasks that are essential to good medical care.

It seems too good to be true, and there's one thing that stands in the way of the successful implementation of Pronovost's checklists – medical arrogance. Some doctors feel that being required to follow such simple procedures implies that they would not take the necessary steps without the checklist and that this is a slur on their competence. Of course, surveys have shown that, indeed, many doctors *do* miss out these steps, but every doctor thinks these lapses apply to other doctors, not to him.

In fact, what Pronovost has done is to bring into medicine an approach that has been tried in the aviation industry, with similar resistance from pilots. In the early days of flying, a number of accidents were caused by the overconfidence of pilots, navigators and engineers. Any suggestion that they might occasionally neglect or forget important safety procedures was taken by flight crews as a slur on their professional competence. But the egos of pilots were deemed to be less important than the safety of passengers and cockpit checklists are now a routine part of every flight. As the news spreads of the beneficial results of Pronovost's checklists, and if doctors and nurses can accept the truth about their occasional lapses, hundreds of thousands of lives and millions of dollars will be saved by a simple tick in a box.

'I LOVE YOU – SHARE MY MHC'

Critics of an exclusively scientific approach to understanding the world often point out that there are things science can't explain – such as love. But it turns out that, when love and attraction determine partner choice, science may well have something to contribute.

Throughout nature, there's a natural avoidance of inbreeding. If relatives mate there can be increased dangers from their genetic closeness. The effect of genes that cause disease or abnormality may be increased if offspring have two copies, one from each parent. Also, if relatives mate, there is less variety in their offspring, which means they may be less likely to cope well in a varying environment. And the offspring of relatives will be more alike and therefore the competition between them will be more intense.

On the other hand, mating with an unrelated partner also has disadvantages. With little or no genetic overlap, some of the useful genes and groups of genes that have built up in a particular family line are dispersed and the advantages lost. Unrelated partners carry new pathogens that could cause infection. Travelling outside the territory of your family can be dangerous and costly. And parenting may be less successful if your partner has acquired different habits in a different environment.

In fact, the optimum mating strategy all the way through the plant and animal kingdom seems to be to choose a mate who differs genetically from you but not too much. Interestingly, the same principle seems to apply to aesthetic judgements – we find most attractive those objects that differ from a familiar standard but not too much. In a recent study of the relationship between kinship and fertility in human couples, it turned out that the greatest reproductive success is in couples who are third or fourth cousins, not closer relatives or unrelated partners.

But how do we make those judgements of genetic similarity? Appearance, certainly – a recent study of partners in a close relationship showed that men were more likely to pair up with women whose bone structure was similar to that of their own mothers and women chose partners who resembled their father in the same way. Some couples look like brother and sister even when they are not. But recent research shows an astonishing ability that we are all unaware of – the ability to detect by smell a specific molecular marker that indicates genetic similarity or difference.

On the surface of white blood cells are molecules that enable the body's immune system to recognize invaders such as bacteria or viruses. The more diverse these molecules the wider is the range of defences against invading pathogens. The information to build these molecules is carried on a group of genes called the 'major histocompatibility complex', or MHC.

So one effect of mating with someone with a very different MHC from your own is to produce offspring with a greater range of MHC genes and therefore a stronger immune system. A more general benefit of being attracted to someone with a different MHC complex is to avoid inbreeding that is too close. A study of couples in isolated religious enclaves showed that couples were less likely to share MHC similarities than would be expected by chance. And a final piece of the jigsaw of love and attraction may be the fact that saliva has a marker for a person's MHC and therefore the process of kissing may be nothing more than an attempt to gauge the MHC status of a potential partner before taking the plunge. It all gives new meaning to the phrase 'kissing cousins'.

TRIBES OF THE INNER ELBOW

Few fields of study might seem more boring than the crook of the human elbow. There are no particular diseases that affect this area, it is very rarely injured, and people certainly don't seek cosmetic surgery for a more beautiful elbow crook. But a group of scientists at the National Human Genome Institute have devoted several years to a close study of this area, presumably on each other, since it would be difficult to carry out detailed research on one's own elbow crook. Their particular interest is the bacteria that live in this out-of-the-way place, and they are part of a bigger project covering the entire human body.

The scientists discovered that there are six distinct tribes of bacteria that live in the inner elbow and that they are quite different from other tribes that live a few inches away on the inner forearm, the subject of study by a whole other team. What's more, however hard you try, you cannot get rid of these bacteria. Even after a thorough wash, there are still a million bacteria clinging tenaciously to every square centimetre of the inner elbow. These bacteria have a job to do and they are not going to give up easily just because we choose to have a shower. Their task, apparently, is to act as skin moisturizers, processing the raw fats we eat and excrete through the skin. They are an example of how essential bacteria are to human life. We tend to hear of them only as causes of disease and targets for antibiotics, but as the National Human Genome research emphasizes, we cannot live without them.

An idea of how many different tasks they perform comes from the discovery that collectively the bacteria on the human body have one hundred times as many active genes as a human. Since most of the genes have a biological function, and presumably the bacteria need only a few genes to carry out their own survival activities, there must be a huge number of functions they perform for us as they go about their daily lives.

It turns out that all over the body there are specialized groups that aren't found anywhere else. There are seventy 'tribes' of bacteria in the body, and although you might expect quite a lot of them to be in the digestive tract, where bacteria play a vital part in our digestion, there are in fact only two tribes of digestive bacteria. The other sixty-eight or so have colonized many other areas in order to carry out specific chemical tasks.

And just to make us feel small, each of us gives house room to twenty times as many bacteria as we have cells in our bodies. Rather than bacteria living on us, we might as well describe ourselves as living on our bacteria.

WHY IS DNA LIKE A KNITTING PATTERN?

Here is an extract from a knitting pattern:

> STITCH COUNT 18, OR 6 PER NEEDLE. ONE CAST-ON ROW OF BLUE, THEN 5 ROWS OF ALL BLUE, THEN AS FOLLOWS:
>
> 6TH ROW: 2 YELLOW, 3 BLUE, 2 YELLOW, 2 BLUE
>
> 7TH ROW: 3 YELLOW, 2 BLUE, 3 YELLOW, 1 BLUE
>
> 8TH ROW: 4 YELLOW, 1 BLUE, 4 YELLOW
>
> THEN THE 9TH THROUGH 13TH ROWS ALL YELLOW, THEN A BINDING OFF ROW OF YELLOW. ONE CAST-ON ROW OF ORANGE, THEN 5 ROWS OF ORANGE, THEN AS FOLLOWS:

> 6TH ROW: I GREEN, 3 ORANGE, I GREEN, 3 ORANGE, I GREEN
>
> 7TH ROW: 2 GREEN, I ORANGE, 3 GREEN, I ORANGE, 2 GREEN
>
> THEN THE 8TH THROUGH 13TH ROWS ALL GREEN, THEN A BINDING OFF ROW OF GREEN.

Anyone familiar with knitting will recognize that this is a series of instructions. If they are carried out in order, long pieces of wool will be turned into an intricate, patterned three-dimensional shape. Knitting patterns exist for every conceivable item of clothing – stuffed toy, tea cosy or scarf – but they all work in the same way – the knitter follows the instructions item by item, line by line, until she (usually) holds in her hand a finished object.

DNA, a substance found in every living cell, is a very long molecule consisting of a series of 'instructions' designed to be read from one end to the other, just like the knitting pattern. Below is a line of of 'instructions' from a DNA molecule. It consists of triplets of letters each of which 'attracts' one of the small molecules in the line below, indicated by the short form of their names. So the three letters GCG, which are strung along the DNA helix, attract a molecule of alanine; CTG attracts leucine, and so on. (You'll notice later in the top line that alanine is also attracted to GCA and leucine to CTA, but that's for another day.)

```
GCG-CTG-GGG-ACG-GGC-GGT-GTT-GGA-GCA-GAG-CTC-TGC-AAT-TTC-TGC-CAA-

Ala-Leu-Gly-Thr-Gly-Gly-Val-Gly-Ala-Glu-Leu-Cys-Asn-Phe-Cys-Gln-
```

If the instructions on one strand of DNA are carried out in order, they will lead to a sequence of small molecules, known as amino acids, joined together, as in the second line above. This sequence is equivalent to the garment that emerges from the balls of wool that are the material acted on by the knitting pattern. The object described by the DNA will be one three-dimensional component of the organism's living processes – a hormone, an antibody, an enzyme, or one of the millions of other molecules that carry out the day-to-day tasks of staying alive.

As with the knitting pattern, it is not clear from just looking at the order of instructions what it is that will emerge after they have been carried out. You have to follow them to find out. In fact, of course, it's a little more complicated than this. The information in the DNA produces a very long molecule called a protein, made up of linked amino acids, and depending on the pattern of the DNA that gave rise to it, it's the protein which turns into a three-dimensional shape, in a way that scientists don't yet fully understand. To push the knitting analogy to its limits, it's as if the knitting process put kinks into a long strand of wool that then assembles itself, before your eyes, into a pair of gloves, a bobble hat, a multicoloured scarf or, eventually, a baby.

A simple protein molecule of a hundred or so amino acids could organize itself into a huge number of possible shapes, only one of which will perform the job required in the body. If this simple molecule tried every single possible shape and took a ten trillionth of a second for each, it would take longer than the life of the universe to try them all, yet scientists don't yet know how it forms itself in an instant into the correct shape.

And the knitting pattern at the beginning of this essay? It's actually a pattern for knitting 'Baby's First DNA Molecule', designed by an enterprising American knitter called Kimberley Chapman. An example of art imitating life.

SOME UNHEALTHY WORDS

If you had to visit a town – let's call it Slagthorpe – that was attacked by an outbreak of the plague, would you prefer a situation in which the prevalence was high and the incidence low, or vice versa? (Obviously, given a free choice, you'd choose one where both were zero.) Often, we use technical words quite untechnically, but prevalence and incidence have very specific meanings in epidemiology.

Prevalence is the proportion of people in a population who have a condition at any one time. So if a disease is very prevalent, there's 'a lot of it about'.

Incidence refers to the rate at which new cases occur in a population during a specified period, expressed usually as cases per 10,000 people per year.

So if you're told that the prevalence of the plague in Slagthorpe is 1 per cent it might not sound too bad, until you find that the incidence is 5,000 cases per 10,000 inhabitants per year. If such an artificial situation existed, it would mean that the plague had only just arrived, but that it was about to spread like wildfire.

If you heard that the prevalence was 90 per cent and the incidence was 1/10,000 per year, it probably means that over a very long period the disease has been gradually extending to most of the population but that it's not very infectious. So go for it, unless you were just planning to drop in for a half a pint and a pork pie, in which case it might be better to go to the next town where there's no risk at all (apart from botulism in the pork pie).

If you did venture into Slagthorpe, what *symptoms* of the plague would you look out for? And what *signs*? People sometimes mistake the meaning of the two words. For example, a leading website for British teachers defines 'symptoms' wrongly as 'the signs of having an illness'. You might experience many *symptoms*, including feeling feverish and noticing swellings under your arms, but you are unlikely to be aware of any *signs* of the disease, since those are defined as objective facts or qualities that are detected by a doctor, such as blood pressure or heart arrhythmia or cholesterol deposits in the surface of the eyeball. Then there's the word *indication* which doctors use, which doesn't mean either sign or symptom, but means a valid reason for using a treatment.

The very thought of contracting the plague might cause hypertension – or is it hypotension? *Hyper* and *hypo* have entirely opposite meanings. Hyper means above normal, and hypo means below normal. But they have the advantage that if you're not sure which meaning is relevant you can slur the end of the prefix, and hope that your listener knows what you are talking about. But the difference can be important.

Hy*per*tension – raised blood pressure – leads to heart disease and strokes; hy*po*tension just makes you faint.

A hy*per*dermic syringe would be very ineffective since it would be squirting its contents above your skin; a hy*po*dermic needle gets beneath the skin where it should be.

And what about *-itis* and *-osis*? Did the doctor say you had 'diverticulitis' or 'diverticulosis'? And what's the difference? Well, 'itis' means 'inflammation of' while 'osis' means a more general diseased condition of the part in question.

THE EYES DON'T HAVE IT

If you were told that a medical diagnostic device correctly diagnosed 88 per cent of patients with kidney disease, you might be convinced that it was worth including in the equipment of the average hospital or doctor's surgery. If you were told that a method of diagnosis that cost nothing at all had the same success rate, you'd probably wonder why it wasn't used by all doctors.

Such a method does exist. It's called iridology, and was discovered in 1836 when a ten-year-old Hungarian boy, Ignatz von Peczely, accidentally broke the leg of his pet owl, and then noticed that it had acquired a black line in its iris. He reasoned that the iris represented the whole body in miniature, and therefore that you could diagnose disease by looking in people's eyes, a practice he then perfected when he grew up to be a doctor.

Like many complementary therapies, iridology is not usually subjected to a double-blind trial in the way that is increasingly the case with orthodox treatments. But dramatic figures such as the 88 per cent claim for kidney disease prompted US medical researchers to investigate further. They presented photographs of the irises of 143 patients, forty-eight of whom had kidney disease, to three ophthalmologists and three iridologists, and asked them to pick out the kidney patients. The results were no better than chance.

Where, then, had the 88 per cent figure come from? Was it wrong? In fact, it was perfectly correct, in the sense that the iridologist who had published those results had looked in the irises of a number of patients, some with kidney disease, and correctly diagnosed 88 per cent of them. The flaw in the study was that he had also diagnosed 88 per cent of the remaining, healthy, patients as having kidney disease when they didn't. The only surprise, perhaps, is that he didn't aim for a 100 per cent success rate by diagnosing *all* his patients, healthy or sick, as having kidney disease, in which case he would certainly have correctly diagnosed 100 per cent of those with defective kidneys.

It reminds me of the old proverb: 'Even a clock that has stopped completely is correct twice a day.'

HOW TO LIVE TO 110 (100 IS PASSÉ)

Imagine having several 100-year-old children, eighty-year-old grandchildren and sixty-year-old great-grandchildren. That is the prospect facing the very small but predictable proportion of us who will live to 120 years old. More and more people are living past the age of 100 – seven in a thousand at the moment – and much of this can be put down to improved health, reduced accidents, better medical treatments and so on. But among the centenarians is a much smaller number who live on and on beyond 110, with a few of those living beyond 120. Scientists who study ageing think that these are not just people who've managed to avoid the diseases or accidents that kill the rest of us but that they are protected in some way by their genes.

In 2006, a survey of thirty-two people between the ages of 110 and 119 showed that although they had some of the conditions of ageing, such as osteoporosis and cataracts, more than 40 per cent of them were still looking after themselves or needed minimal help with daily tasks. What was significant about them was that they showed a lower proportion of the killer diseases of old age – strokes and heart disease, diabetes and Parkinson's disease. What's more, when super-centenarians, as they are called, did eventually die, post-mortems showed that it was not from cancer, heart disease, stroke or Alzheimer's but from a disease that is extremely rare in younger people (ninety- and 100-year-olds). This disease is called 'senile cardiac TTR amyloidosis'. It's a type of furring of the arties like atherosclerosis, but with an entirely different cause. The arteries become clogged by a protein called TTR, which is connected with the thyroid hormone, thyroxine.

The genetic basis of superageing became clearer when scientists looked at the relatives of the supercentenarians. In a comparison of the longevity of siblings of 110-year-olds and older, the brothers lived twelve to fourteen years longer than other men born in the same year, and the sisters gained eight to ten years.

One question scientists are still trying to answer is whether the odds of dying in a particular year – which increase relentlessly as we get older – actually plateau out for the very oldest, so that once we've reached a very advanced age we may be no more likely to die in the following year, or perhaps even less likely. Perhaps that means that a cure for amyloidosis would lead to a whole new generation of people whose bodies have no real excuse for dying.

PREMATURE PEEPSHOW

In 1897, a German physician called Martin Couney travelled from Paris to London with three wicker baskets full of French premature babies, to be put on show at the 'Victorian Era Exhibition' at Earls Court in London. Couney was exhibiting an incubator for premature infants, as part of an exhibition of new medical technology, and he felt the display would have more impact if the incubators contained live babies. He had done this with no harm to the babies at a German exhibition, where the display was called a 'child hatchery', but for understandable reasons British doctors had refused to allow British premature babies to be displayed in this way, which is why a group of Parisian babies were subjected to long train journeys and a Channel crossing to become a sideshow in the Victorian exhibition.

The show was a great success, with 3,600 visitors on a single day, but in the words of a *Lancet* editorial: 'The success . . . has proved a mixed blessing. It attracted the cupidity of public showmen, and all sorts of persons, who had no knowledge of the intricate scientific problem involved, started to organise baby incubator shows just as they might have exhibited marionettes, fat women, or any sort of catch-penny monstrosity.'

There was rash of imitators who attempted to borrow premature babies from hospitals around the country, and Couney had to write a letter to the *Lancet* warning that he ran the only authentic 'Infant Incubator Institution'. Nevertheless, as the *Lancet* pointed out,

> at the World's Fair held at the Agricultural Hall, Islington, there is an incubator show where the charge for admission is only 2d. We fail to see how this small sum can cover the cost of properly trained attendants and of wet-nurses. . . . The infants breathe the atmosphere of the interior of the Agricultural Hall, where, apart from the numerous visitors, the whole of Wombwell's menagerie is kept. Just opposite the incubators there are some leopards and everyone is familiar with the obnoxious odour that arises from the cages in which such animals are incarcerated. . . At Barnum and Bailey's Show also there is a baby incubator show . . . but . . . what connection is there between the serious matter of saving human lives and the bearded woman, the dog-faced man, the

> elephants, and performing horses and pigs, and the
> clowns and acrobats that constitute the chief attraction to
> Olympia. . .?

None of this criticism had any influence on Couney's activities, and he continued to show up at exhibitions and fairs
around the world. In 1901, his exhibit received rave reviews at
the Pan-American Exposition in Buffalo, not far from the
Niagara Falls, which led one hyperbolic commentator to write:

> Two vast extremes. The falls of Niagara with the great
> system of lakes and rivers behind them; the diminutive
> baby in its hot-air chamber, sightless, deaf, feeble – but
> with the great human race, the vast sea of organised
> thought back of it. . . What is the power of the falls beside
> the force that may originate in the tiny brain of an incuba-
> tor baby? The brain is smaller now than half an apple. But
> that brain may start a work that will persist and affect
> man's destiny when the falls shall have dwindled down to
> an even placid stream. . .

But three years later the incubator-exhibiting business
received a setback, when babies put on show by one of Couney's
competitors at the Louisiana Purchase Exposition in St Louis
suffered an epidemic of diarrhoea, killing about 50 per cent of
the babies.

No more competitors were brave enough to take up the challenge, particularly when it cost the exhibitor about $15 a day per
baby, and the field was left open to Couney, who was now living
in America, and he organized an annual incubator exhibit on
Coney Island. (Oddly, the original spelling of Couney's name
was Coney and his brothers still kept that spelling.) Couney's

knowledge of incubators came in handy when his wife, an incubator nurse, gave birth prematurely and their daughter spent the first three months of her life in one of his exhibit incubators.

Couney showed his incubators at several more large exhibitions, including the New York World's Fair in 1939. Forty years of experience had adapted and improved the incubators themselves and the routine for caring for the babies, and these advances were now commonplace in major maternity hospitals. But while the physical health of the babies was handled well, neither Couney nor anyone else thought much about the psychological condition of tiny premature babies and their mothers deprived of contact for many weeks. Parents were allowed in free, while ordinary visitors paid 25 cents, but since the babies came from a wide radius around the exhibition site, parents were not always frequent visitors. In fact, Couney was puzzled by the fact that sometimes when it came time to hand the babies back to the parents, they were reluctant to resume their parental responsibilities.

This odd marriage of showbiz and medical research probably did benefit the embryonic science of neonatology more than it harmed it, and Couney himself was recognized as a pioneer in the field. There is even a plaque on the wall of a Holiday Inn in Atlantic City to mark the site of one of his old shows.

MANY COAGULATIONS

Most of us have scars on our bodies, reminders of occasions when the skin was breached by something sharp or pointed. If our bodies were working normally, blood would have flowed for

209

a short while before a clot formed, which stayed in place while the skin grew underneath it. Then when the skin was whole again, the clot, now a scab, fell off, and left a small light area of skin which may be with us for ever.

The way in which the body deals with injury is one of many physiological processes that turn out to be both familiar – we've all cut ourselves and waited for it to heal – and almost unbelievably complex. Through long millennia of evolution, and much trial and error, animal bodies have devised mechanisms for dealing with injury which now seem exquisitely suited for their purpose but, like all evolution by natural selection, were based on the early death of creatures that lacked the ability to close up their blood vessels and make new skin when they were injured.

The process that stops all our blood draining away when a blood vessel is cut has four stages. First, the body tries to constrict the blood vessel as much as possible to reduce the size of any gap through which blood could escape. Second, a collection of blood components called platelets, smaller than red blood cells, gathers at the injury site to form a loose plug. This activity is triggered by the lining of the blood vessel that is exposed by the injury, and normally protected from contact with the blood flow. Third, since the plug of platelets is only temporary, work begins on constructing a fibrous mesh or clot which will trap the plug and keep it in position. Fourth, at some point, while the tissue around the wound is being repaired, the clot must be dissolved to allow the blood to flow freely through the artery or vein.

Every one of these steps is achieved through a series of smaller steps which use molecules already in the blood or called up specially for the purpose. Each of these molecules has a chemical name which trips off the tongue of biologists but is bewildering and uninformative to the rest of us. To give an idea of the complexity of just one stage, the initial clotting process, I'll give these chemical names letters.

As I said, the clotting process starts when the wall of the blood vessel is breached. The damaged cells expose chemical A to the bloodstream. Chemical A binds to another chemical, BC, which is actually a combination of two proteins, B and C. B activates D, which in turn converts chemical E to F. F does several different things, including dividing G into two types of molecule, H and I, and activating J, which converts the soluble fibres in the blood to an insoluble mesh that holds the clot together. Meanwhile, we left C doing nothing, but not for long. C's job is to magnify some of the above events so that the clot is formed more quickly. So C binds to K which is stabilized by another chemical, L. C and K together then activate more B (which can then do more of what it was doing before). At the same time, F activates more D, and K, as well as a new player, M, and M amplifies the production of C.

This may be confusing, but probably not as confusing as if I'd given the components their real names, which are:

A   – tissue factor
BC – Factor VII
B   – Factor X
C   – Factor IX
D   – Factor V, aka prothrombinase
E   – Factor II, aka prothrombin
F   – thrombin
G   – Factor I, aka fibrinogen
H   – fibrin
I   – fibrinopeptides
J   – Factor XIII

K – Factor VIII
L – von Willebrand's Factor
M – Factor XI

Every one of these factors is essential for successful healing, and if any of them is absent, perhaps for genetic reasons, it can lead to a disease of clotting. One of the most common, von Willebrand's disease, is due to an absence of the chemical I've called L above, von Willebrand's factor. A deficiency of the chemicals I've called K or C or M can lead to different types of haemophilia, where a minor cut can lead to major and sometimes fatal bleeding.

Almost as amazing as the cascade of events that accompanies a simple injury is the ingenuity of the biologists over the years who have teased out the details of this process by painstaking experiment and chemical analysis.

'RADIATION IS GOOD FOR YOU'

No more drastic 'test' of the effect of radiation on human beings can be imagined than the explosion of an atomic weapon over a major city. With much discussion nowadays of the effects of radiation that might be expected from accidents at nuclear power stations or the disposal of nuclear waste, people have returned time and again to the devastating effects of the atomic bombs dropped by the Allies over Japan in 1945, as the major – inadvertent – source of facts about how the human body is affected by doses of radiation.

While no one doubts the appalling effects of a blast of radiation on a person within a few miles of the explosion, there has always been a question that needed answering about the possible effects on future generations as a result of either parent being affected by high levels of radiation. It was suspected that sperm or ovum cells might receive doses that caused mutations leading to defects in offspring.

To test this idea, a US commission was set up in Hiroshima to monitor the children of survivors to see whether they developed a greater number of abnormalities than equivalent populations in other parts of Japan that had not been exposed to radiation. For forty years, until all the survivors were past childbearing age, detailed data were gathered about the health of their children and grandchildren. About 75,000 people were followed, from childhood to adulthood, and the researchers looked not only at disabilities and deformities, but also diseases like leukaemia that are known to be connected in some way with radiation.

The researchers were looking for an *excess* of cases between the two populations. If you take a congenital condition like spina bifida, the scientists knew of, and confirmed, an incidence of about 1 per cent in the general population. For there to be a radiation effect, therefore, they would be looking for a higher proportion in the children of atomic survivors.

The researchers failed to find any difference in the incidence of spina bifida between the survivors and the unexposed subjects. And they failed too with all the other conditions they thought might be a result of radiation affecting the reproductive systems of atomic survivors.

So do the figures say that radiation is not bad for you? Bizarrely, taken at face value, the researchers found something even more surprising than that. Their analysis showed that the children born to atomic bomb survivors seemed to be *healthier* than children of the same age born elsewhere in Japan. There were fewer stillbirths, a smaller numbers of deaths from cancer, and lower overall mortality.

As with so many scientific problems, while people hope for a clear and simple answer they don't always get it. No one knows why the researchers found the results they did, and there have been plenty of people who cannot believe the figures and who have tried to find alternative explanations for the data. But for the time being, while we can't actually say 'radiation is good for you' what we can say is that in this one area – the effect of a large dose on the descendants of a survivor – there is no evidence that it is harmful.

THE HAIR OF THE DOG

One of the most unpleasant symptoms of a hangover (so I've heard) is the vertigo that can occur when you try to stand up and walk around. The sense of balance depends on sense organs in the inner ear which are thrown out of kilter as the level of alcohol in the blood rises. The way in which this unpleasant feeling changes in the hours after drinking is a result of a subtle interplay of factors, and has given rise to an odd observation which has probably been noticed since the Babylonians first drank date-palm wine.

The organs of balance are three semicircular 'canals' roughly at right angles to each other filled with fluid. One 'canal' detects rotation of the head around a vertical axis, another detects 'nodding' movements, and the third detects the rotation of the head around an axis from the nose to the back of the head, the sort of movement we've all experienced – I'm sure – when cartwheeling.

In normal alcohol-free life, the density of the fluid in the canals is similar to that of the blood. As you move your head the fluid flows back and forth in the canals and pushes against pressure sensors, which send signals to the brain. The combined signals from all three canals form an ever-changing 'GPS' signal allowing us to know where the head is in space at any time.

But an important aspect of the system is that, for accurate sense messages to flow from the organs of balance to the brain, the density of the fluid in the canals should be the same as the density of the blood. If the relative densities between blood and inner ear fluid change, the system makes the wrong judgement about the position of the head, a judgement that is contradicted by the message we receive from our eyes. And drinking alcohol changes the relative density of the fluid inside and outside the semicircular canals leading to conflicting messages and a sense of confusion.

As alcohol levels rise, the blood becomes *diluted* relative to the fluid in the inner ear, because alcohol is less dense than blood. This difference causes the first phase of dizziness and nausea. But then the alcohol seeps into the semicircular canals and corrects the imbalance, so there's a period when we feel normal again. After we stop drinking, the alcohol in the blood is gradually broken down by the liver, and the blood density rises to its normal level. But now we have another problem, because the fluid in the inner ear still contains alcohol, which has not been broken down, and so the blood is now *denser* compared with the fluid in the canals. Just as there was a time lag before the alcohol seeped into the organs of balance, there is now another time lag until the alcohol seeps out and the fluid is back to normal density. During that time lag, we feel symptoms of vertigo again, but in the opposite direction from the first post-drinking phase.

'When will it all end?' we think. But hope is in sight. The Hungarian phrase '*kutya harapást szörével*' tells us why. This means 'you may cure a dog's bite with its fur', or, as we say in English, 'the hair of the dog that bit you' or, in French, '*rallumer la chaudière*' – relight the boiler– or, in Danish, '*rejse sig ved det træ, hvor man er faldet*' – you should get up next to the tree where you fell. (Hangovers are clearly a global – or at least a European – problem.)

Now that we understand that it is the imbalance between the density of the blood and the density of the fluid in the inner ear which causes the vertigo and nausea, we can see why the hair of the dog works as hangover relief. On the morning after the night before, when the blood is back to normal, the inner ear fluid is still diluted by alcohol because of the slower seepage of the alcohol out of the canals. But if we take another drink, the blood becomes diluted too, and the balance between the two fluids is restored.

Cheers.

EUREKA ON HIGHWAY 128

Taking tiny samples of DNA from a crime scene and comparing them with DNA from a suspect depends on being able to make millions of copies of the initial scrap of DNA so that there is enough material to work with. This duplication process depends in turn on a scientific technique called PCR, standing for polymerase chain reaction, invented by one of the world's odder scientists, Kary Mullis, who won a Nobel Prize for his discovery.

Each Nobel Prize winner is expected to give a lecture about his work and in his lecture Mullis described how the idea for

PCR came to him in the middle of the night when he was driving with his girlfriend to his cabin in the woods to the north of San Francisco. In the course of his lecture we learn of his girlfriend's name – Jennifer Barnett – and even how Mullis had strayed from his wife, Cynthia, to spend two tumultuous years with Jennifer.

On this particular night, while driving up Highway 128, Mullis was wrestling with the problem of how to analyse the structure of a specific piece of DNA when you only have a tiny sample. Techniques existed for working out the particular sequence of units – the genetic fingerprint – of a whole batch of identical DNA, but people on the whole don't leave behind buckets of blood or other DNA-containing fluids, unless of course they're dead. What happened more often was that a tiny sample of blood or sperm or saliva would be discovered which was far too small for any useful analysis. But Mullis suddenly thought of a way of magnifying the tiny sample of DNA many millions of times, turning it into a large and analysable quantity. In his Nobel lecture he described the circumstances of his prize-winning insight:

'As I drove through the mountains that night,' Mullis said, 'the stalks of the California buckeyes heavily in blossom leaned over into the road. The air was moist and cool and filled with their heady aroma. My girlfriend Jennifer was asleep.'

DNA is normally two long intertwined strands carrying the unique genetic description of an organism [🐕 199]. Mullis suddenly conceived of a method of separating the two strands and attaching markers to each end, and then using another molecule, called DNA polymerase, to make a copy of the DNA between the markers, thus doubling the amount. As he thought about this he realized that by repeating the process over and over again, he could produce huge amounts of identical DNA.

'EUREKA!!!!' he said in his Nobel lecture.

> I stopped the car at mile marker 46.7 on Highway 128. In the glove compartment I found some paper and a pen. I confirmed that two to the tenth power was about a thousand and that two to the twentieth power was about a million, and that two to the thirtieth power was around a billion, close to the number of base pairs in the human genome. Once I had cycled this reaction thirty times I would be able to obtain the sequence of a sample with an immense signal and almost no background. 'Dear Thor!' I exclaimed. I had solved the most annoying problem in DNA chemistry in a single lightning bolt.

Mullis couldn't wait to tell his sleeping girlfriend. 'Jennifer, wake up,' he said. 'I've thought of something incredible.' But Jennifer failed to respond to this Great Moment in Chemistry.

We are not told whether Jennifer's age was 27 years and 142 days, but it is quite likely that it was, or had been at some point during their acquaintance, since Mullis's eccentricities are said to include a liking for women of precisely 10,000 days old, some of whom appear as slides in his chemistry lectures, clad only in multicoloured fractal patterns.

A LA RECHERCHE DU PONG PERDU

There are very few similarities between my genes and Marcel Proust's, as far as I am aware. We certainly have no common ancestors, unless you count the fact that he is Jewish and my father was Arab, and both ethnic groups are believed by legend to descend from Noah's son, Shem, hence being Semites. But we do have one genetic trait in common, Marcel and I – a hereditarily determined hypersensitivity of smell for particular chemicals that are excreted in the urine of someone who has just eaten asparagus.

I know that Proust had this characteristic because in the first volume of *A La Recherche du Temps Perdu* he includes a paean to asparagus which describes its subtleties of colouring and finishes up with the following:

'I felt that these celestial hues indicated the presence of exqui-site creatures who . . . allowed me to discern . . . that precious quality which I should recognise again when . . . they played (lyrical and coarse in their jesting as the fairies in Shakespeare's *Dream*) at transforming my chamber pot into a vase of aromatic perfume.' Another great writer with a penchant for the smell of asparagus-scented urine is Gabriel Garcia Marquez, who writes of a character in his book *Love in the Time of Cholera* taking 'pleasure in the vapours of his own fragrant urine.'

Well, we know Proust and Marquez to be men of refined tastes, so I can't explain why they described methylmercaptan, a similar chemical to that which gives a skunk its smell, as 'aromatic perfume' or 'fragrant'. But it is one of the chemicals that imparts a characteristic smell to the urine of people who eat asparagus.

Of course, what I've said so far may not make much sense to the majority of you who are not familiar with the phenomenon I'm talking about. What's more, if you are in that group you will *never* have the experience that Marcel and I have shared, because you are congenitally incapable of smelling the by-products of asparagus in your urine.

Considering how long asparagus has been seen as a delicacy – the Romans sent asparagus fleets round the Mediterranean to gather it – it's surprising that it was only in 1980 that a definitive experiment was carried out to determine exactly what was going on.

There were several possibilities:

1. Some people excrete smelly asparagus by-products and can detect them in the urine; others don't and can't. For a time it was believed that there were people with a type of genetic effect called an inborn error of metabolism who were unable to process the constituents of asparagus and therefore excreted molecules, such as methylmercaptan, that other people were able to break down in the body.

2. Some people excrete and don't detect; others don't excrete and do detect; a third group excretes and detects.

3. All people excrete, but only some people detect.

It doesn't take much more than the use of a shared bathroom after a romantic asparagus supper to narrow down the options. A proportion of couples living together will know that it's possible for one to detect the odour in the urine of both, while the other hasn't the faintest idea what his partner is talking about. A survey in the UK of 800 people, male and female, showed that about 40 per cent produced the odorous chemicals after eating asparagus, and this seemed to be genetically determined. However, in a survey carried out in France, 100 per cent of the subjects produced the chemical in their urine after eating asparagus. Proust himself would have been proud of his fellow-citizens' superiority over British urinators.

But what about the ability to smell the chemical? This too seems to be genetically determined. The definitive experiment was carried out in 1980 by three Israeli doctors, and it doesn't make pleasant reading, so I'll abbreviate it. A man who'd eaten 450 grams of canned asparagus supplied a urine sample which was divided up into smaller samples of different dilutions, from no urine present to the real McCoy at full strength. A range of patients were recruited from an outpatients clinic and asked to sniff the samples. The researchers found that a small group, about 10 per cent, were 'smellers' who could detect the offending chemical at quite low dilutions, and the rest were 'non-smellers' who couldn't detect it at all.

Scientists already knew about certain types of 'taste blindness'. There is a chemical called phenylthiocarbamide (PTC) which tastes bitter to about 75 per cent of the population and is tasteless to the other 25 per cent. But no one had come across proven examples of 'smell blindness' until the Israel doctors' experiments in 1980. What the researchers believe is that there is a specific genetic trait in a minority of the population which leads to them being sensitive to the offending molecule. Of course, none of this explains why Proust found his chamber pot filled with 'aromatic perfume' while mine, if I ever used one, would be worth emptying as quickly as possible.

SEX AND SCIENCE

There are many important issues yet to be resolved in the interaction of gender and science. Many feminists, and others too, think that there are not enough women in science and that the education and selection of scientists is skewed against girls and women. Perhaps this leads to more emphasis being given in restricting research topics to the things that men like to study as opposed to women's favourite topics – always assuming that such a difference actually exists.

It is certainly true that the majority of scientific discoveries have been made by men, for fairly obvious social reasons, but could it be that the actual content of science as we believe it today has been shaped by that fact? Might we now believe different things about the universe and how it worked if women had been carrying out the research and having the ideas?

There is a small but significant group of philosophers and sociologists who believe this is the case. Take one of the best-known and now best-established equations in science – $E = Mc^2$. A Belgian philosopher and feminist, Luce Irigaray, has written: 'Is $E = Mc^2$ a sexed equation? Perhaps it is. Let us make the hypothesis that it is insofar as it privileges the speed of light over other speeds that are vitally necessary to us. What seems to me to indicate the possibly sexed nature of the equation is not directly its uses by nuclear weapons, rather it is having privileged what goes the fastest. . .'

She seems to be suggesting that because men are believed to be interested in things that go fast, and, indeed, the faster the better, this important equation is expressed in terms of the speed of light rather than other 'speeds that are necessary to us', although I can't think of another speed that is somehow more feminine at the same time as being an important physical constant.

Feminist theorists also complain about how biologists use male rats in studying animal behaviour and physiology. One has written: 'The implicit assumption is, of course, that the male rat represents the species.' Another points out that researchers into fertilization describe the female egg as 'drifting' or sitting passively, while the macho sperm 'burrows' and 'penetrates'. As a man, I find it difficult to think of more feminine or gender-neutral words to describe what happens as a sperm cell travels at speed ('speed' again) and meets a much slower or stationary ovum before finding itself beneath the surface when a moment before it was outside.

The language of physics comes under attack as being too masculine, because it speaks of atomic particles 'bombarding' samples, and 'colliding' with other particles. Asked what language should be used instead, for what is a basic technique in particle physics, one feminist has written that colliding atoms should instead be described as 'joining together for mutual benefit'.

A topic developed at some length by Irigaray, and supported by other feminists, is the field of fluid dynamics. This is the scientific study of the motions of fluids, covering topics such as turbulence, viscosity, flow and shock waves. Much of this branch of physics is mathematical, using differential equations and computer simulations.

Irigaray believes that fluid dynamics is a *feminine* science and is therefore treated as a poor relative among the physical sciences. (She is not a scientist herself and has been criticized for quoting no texts or research in the field.) She believes that by contrast solid-state physics is masculine and therefore privileged and that a fluid-dynamic topic like turbulence is not well understood because it is of less interest to men. Another feminist writer, Katherine Hayles, has summarized Irigaray's ideas as follows:

> Whereas men have sex organs that protrude and become rigid, women have openings that leak menstrual blood and vaginal fluids. Although men, too, flow on occasion – when semen is emitted, for example – this aspect of their sexuality is not emphasized. It is the rigidity of the

> male organ that counts, not its complicity in fluid flow . . .
> In the same way that women are erased within masculin-
> ist theories and language, existing only as not-men, so
> fluids have been erased from science, existing only as not-
> solids.

Does any of this make sense, and in any case does it matter? It is surely laudable to draw attention to aspects of a major activity of society, the pursuit of scientific knowledge, if the way it is carried out disadvantages women. But for me, taken to extremes like these, it is an attack on the validity of the scientific technique itself. If women scientists, given their freedom, would make different and even contradictory discoveries and deductions about how the world works, what is the use of science as a tool for understanding the universe? And yet, this almost laughable view is what some philosophers believe. For Hayles and others, the equations of fluid dynamics, which underlie the success of every flight that takes off and lands at an airport, are not necessarily correct. With more women in the field, 'people living in different kinds of bodies and identifying with different gender constructions might well have arrived at different models for flow'. If they did, I suggest, they would not have been very good scientists.

WHICH WOMAN HAS MADE THE GREATEST CONTRIBUTION TO MEDICAL RESEARCH?

There have been – and are – many brilliant medical researchers who are women. But Henrietta Lacks is not one of them. As a mother of five children, and living in a suburb of Baltimore in

1951, she had no connection with medical science, although she lived within three miles of one of America's leading medical research institutes, at Johns Hopkins University. But the unfortunate development of a cervical tumour after she had given birth to her fifth child started an extraordinary chain of events. She went into Johns Hopkins hospital for treatment and some of the cells in the tumour were taken – without her permission, which wasn't required in those days – and analysed by a doctor at the university who was studying cancer cells.

The doctor, George Gey, discovered that Henrietta Lacks' cells had a very unusual property. Unlike many human cells, they were able to live indefinitely outside the human body, dividing and redividing without limit. In other words they were immortal, when most cells taken from the body would die out after a certain number of generations, about fifty-two in the case of normal human cells. Gey had been looking for such cells to observe and test in his research programme into the causes and treatment of cancer, and for reasons no one knows, Mrs Lacks' cancer cells fitted the bill. Because her tumour was extremely fast growing, the cells taken from it all showed the same properties of strength and growth. While Gey was testing the cells and beginning to use them for his own research, he realized that they would also be a valuable tool in other areas of medical research, including the study of leukaemia and radiation damage, the investigation of genetic control mechanisms, and research into how cells make protein molecules. Soon Henrietta Lacks' cells were being shipped around the USA and beyond, to Russia, South America and anywhere else where researchers were trying to study mechanisms of human disease.

In one area of research her unusual cells provided a very timely solution to a problem confronting polio researchers. At the time of her illness, America was facing a series of polio epidemics. Scientists knew they needed a vaccine to combat the disease, and Jonas Salk was racing Albert Sabin to come up with an effective one. Since the polio virus multiplies in living cells in the human body, scientists needed a strain of human cells that could be used to grow the virus which would be inactivated and then used for vaccination. But they couldn't use just any old cells – they needed some that would have the identical properties and would reproduce over and over again to provide the large numbers that would be needed once a vaccine was perfected. What's more, such cells would be essential for discovering which strain of polio was affecting people in a particular epidemic, so that the vaccine could be targeted to that strain. Henrietta Lacks' cells fitted the bill. They were some of the strongest cells known to science and reproduced an entire generation every twenty-four hours.

Soon every major research laboratory had samples of the cells. They were known as HeLa cells, and initially the identity of the donor was concealed. They were believed to come from a woman called Helen Lane or Helen Larson. In fact, Mrs Lacks' husband didn't find out about the use of his wife's cells until 1975, more than twenty years after their value to medicine had been determined and at a time when they had proliferated all over the world and even into space as part of NASA's Science in Space programme.

The properties of her cells that were valuable to science made them lethal to Henrietta Lacks. After her cancer was diagnosed, she lived another eight months under the onslaught of the cells, which spread quickly to other organs producing cancers that were impossible to treat, and she died on 4 October 1951.

One scientist has proposed that HeLa cells be defined as a new species, because certain properties of the cells set them apart from any other organism. But the genus *Helacyton* has yet to be generally accepted as having a separate perch on the tree of life.

As an example of what's sometimes called the Law of Unintended Consequences, there's a sting in this tale. So successful were the HeLa cells in replicating that they began to take over the cell collections of some laboratories, contaminating samples, so that when researchers thought they were working with cells from the breast, the prostate or the placenta – and based the conclusions of an experiment on that belief – they later discovered that their samples had been replaced by the HeLa cells and their research was worthless.

ONE BLIND BABY OR SIXTEEN DEAD ONES? THE CHOICE IS YOURS

This stark choice was posed by a British researcher in 1973, after an analysis of one of the most complex issues in twentieth-century medical ethics. It all began with the discovery of a rare type of blindness in a premature baby in 1941, which turned out to be the forerunner of a series of similar cases which rose to epidemic proportions during the 1940s and 1950s. It reached the stage where one in eight premature babies weighing less than four pounds and reared in hospital incubators was going blind.

The blindness was called 'retrolental fibroplasia' or RLF, because its most important characteristic was a membrane covered with distorted blood vessels which grew behind the lens of the eye. By 1953, 10,000 children had been blinded by RLF, 7,000 of them in the USA, and this rapid increase suggested that the condition was caused by some new development in medicine, and in particular in neonatology, the science of caring for newborn babies.

There was a rash of research projects which looked at possible causes. Could it be the misuse of vitamin supplements or hormones? What about the ultraviolet light used to reduce jaundice? Was there something about the incubators themselves? There were several false leads before attention turned to the use of oxygen to improve the survival of very small babies with breathing difficulties. This was widespread in the United States, where most cases of RLF occurred, and not used very much in Britain, where there were far fewer cases.

One way to test this theory was to organize a trial in which researchers gave high levels of oxygen to half of a group of premature babies and low levels to the other half, the control group, to see whether there was any difference in the incidence of RLF. But in the hospital where the trial was carried out, nurses were so convinced that the babies in the control group would be harmed by a shortage of oxygen that they secretly switched the oxygen back to 'high' again. When this was discovered, and the experimental method was changed to avoid such tinkering, the results were dramatic. In the high-oxygen group, seventeen out of twenty-eight infants developed minor or severe RLF; in the low-oxygen group, of thirty-seven babies, only six developed RLF, and in a minor form.

So far, then, a triumph of medical research. The results of this small study were confirmed by a much bigger research project involving nearly six hundred babies in which eighteen hospitals collaborated. Babies needing oxygen for their breathing difficulties were given either routine oxygen or reduced amounts. More than twice as many babies in the higher-oxygen groups showed signs of RLF compared with those given less oxygen.

The main effect of the data from the research studies was to make sure that oxygen levels were kept below 40 per cent, although even then a few cases occurred, some as a result of the low levels of oxygen that occur in normal air.

But of course, in these cases, with babies of very low birth weight, oxygen was not given on a whim. There was evidence that the oxygen helped babies who might otherwise die in the earliest days of life to survive.

What's more, when researchers looked at one particular respiratory condition, called hyaline membrane disease, which diminished with the use of oxygen, they found that as oxygen levels were dropped to prevent blindness from RLF, deaths from hyaline membrane disease went up.

It was this observation which led to the bleak calculation that 'each sighted baby gained may have cost some sixteen deaths'.

Now, more than sixty years after the initial discovery of RLF (now called ROP – retinopathy of prematurity), another multi-centre study, involving research hospitals around the world, is trying to settle definitively whether this equation of horror actually reflects the true situation, and if so, what can be done about it.

The whole story illustrates the dangers of allowing new medical treatments to be introduced just because a doctor or researcher believes that they will work.

As a leading paediatric researcher, William Silverman, wrote:

A doctor may (with impunity) prescribe a 'fashionable' untested treatment because of the advice of an authority or a colleague who is a personal friend; because he has read about it in the newspapers, in advertisements, or in medical journals; or simply because the treatment 'makes good physiologic sense.' On the other hand, if he should decide that there is sufficient uncertainty to warrant caution, and he chooses to undertake a planned test, his action is subject to criticism. 'I need permission to give a new drug to half my patients, but not to give it to them all,' said [one doctor], commenting on the absurdity of this situation.

A 'SUGAR SCALPEL'

We've all heard of a sugar pill as a possible cure for illness, using the well-known placebo effect, which works on the basis that if a patient believes the medicine will work, that belief often helps the patient's condition improve. But the idea of a 'sugar scalpel' – a surgical procedure that has a placebo effect – is a little more difficult to swallow. And yet, in the days when ethical standards in medical research were not so tight, a research team in Seattle decided to see whether there could be a placebo effect from surgery, so that a patient could get better as a result of believing that he'd had a specific type of surgery when he hadn't.

To find out whether a treatment really works is fraught with pitfalls. Clinical trials are essential to sort out genuine effects of specific treatment from the other factors that make patients feel better. The most rigorous type of clinical trial is one called 'double-blind'. Here, patients are randomly allocated to two groups, as evenly matched as possible for characteristics like age and sex and state of health and so on. Then a treatment, usually a drug, is given to one group but not to the other, who receive a placebo or dummy pill.

'Double-blind' means that neither the doctors running the experiment nor the patients know who is getting the treatment and who is getting the placebo. This is because both the patient's and the doctor's knowledge can condition how the patient responds. If you think you're getting a brand-new, innovative treatment you may feel buoyed up and your body could respond positively. Similarly, the doctor's knowledge of who is getting the real treatment may well communicate itself to the patient, or may cloud the doctor's judgement when he assesses which of the patients has improved and which hasn't. So, it's best if doctor and patient are 'blinded' to the secret of who is getting the real treatment.

In 1939, in an effort to treat a common heart symptom, angina, which at the time had fewer effective treatments than today, an Italian surgeon, Davide Fieschi, devised an operation to tie, or 'ligate', two arteries near the heart in the belief that this would divert more blood into the coronary arteries, increasing the oxygen flow and reducing angina. The results were spectacular. After the operation, as many as three-quarters of the patients experienced a considerable decrease in their angina pain, and a third of them were 'cured'. Despite everyone's satisfaction with the result, it was generally acknowledged that no one knew precisely how the operation worked and nobody had conclusively demonstrated that the hoped-for diversion of blood to the heart did really happen. (It was known, for instance, that the operation didn't work when done on dogs.)

So in the late 1950s, a group of researchers in Seattle decided to look at the effects of this operation in greater detail and work out what was going on. The two groups both carried out studies that would probably be forbidden by current ethical guidelines today, but which were perfectly acceptable at the time. Patients with angina were randomly divided into two groups. One group got the standard internal mammary artery ligation operation. The other group, however, got a 'sham' operation in which the chest wall incision was made but the internal mammary artery was left untouched. None of the patients was told that they were in a research study (which would be illegal now) and none of the doctors who assessed the patients after surgery knew which operation had been performed.

The results of these studies were quite clear – and very upsetting for thoracic surgeons. The operation improved the pain, but the improvement had nothing to do with the actual ligation of the internal mammary arteries. In one study, both groups reported the same subjective improvement after surgery – 32 per cent average improvement in the group that had had the real surgery and 42 per cent improvement in the group that had had the sham operation. Both groups also reported the same effect on their need for nitroglycerin tablets, the modern way to reduce the pain of angina. The group that had the real operation decreased their daily consumption by 34 per cent while the 'sham operation' patients reduced their dose by 42 per cent. But despite the fact that these patients felt better, there was very little improvement in objective measurements – the average improvement in the amount of exercise they could perform was only one minute or so.

It was quite clear that the operation was beneficial for reducing angina, but equally clear that the act of tying off the internal mammary artery (the whole point of the operation in the first place) made no contribution to the success of the procedure. Furthermore, it was obvious that patients could feel very much better (with less pain and a decreased need for tablets) but could still have exactly the same objective problems (reduced ability to exercise and abnormal ECGs during exertion). The patients *felt* better but weren't *getting* better, and what made them feel better was 'something that goes with having an operation on the chest', not the tying off of the internal mammary arteries.

In the 1960s and 1970s, the idea of using a placebo form of surgery in controlled trials was seen increasingly as unethical. A sugar pill is going to harm no one, but doing an operation, however minor, carries some risk. There are many surgical procedures for which the evidence of benefit is no greater than in Fieschi's operation for angina, however, and unless some form of clinical trial is done, no one can rule out the placebo effect. In the last ten years or so, cautiously but some would say unethically, researchers have begun to bring back placebo surgery.

In 1999, a research team in the USA reported an experiment to test the effectiveness of a new treatment for Parkinson's disease. The treatment involved drilling holes in the patient's head and injecting fetal cells, which, it was hoped, would repair the patient's brain. This had been shown to reduce the symptoms in some patients. But to be sure that it was the fetal cells that were producing the improvement, rather than the drama and excitement of having holes drilled in their heads, the researchers took forty Parkinson's patients, drilled holes in all of them but injected the cells into only half of them. One year later, although a good proportion of the treated patients had improved, so, too, had three of the twenty patients in the placebo group.

Other trials involving knee surgery and spinal surgery were also carried out, to be sure that new treatments devised on the basis of medical theories actually worked the way they were meant to.

Of course, the question arises: if people feel better after

sham surgery, why not offer it, in the same way that some doctors use sugar pills when they suspect that they will make a patient feel better? But this addresses a whole new area of honesty and trust in the doctor–patient relationship. Some people are already uneasy with a doctor saying to a patient: 'I'm going to give you this oddly shaped and brightly coloured pill called "Uvebinfuledium" and you should feel right as rain,' when he knows it's a placebo. The deception would be all the greater – and much more expensive – if the patient was subjected to the charade of a surgical operation.

BARING YOUR SOLES TO X-RAYS

When I go to the dentist to have a dental X-ray, I am fitted with a lead collar and the dentist and her assistant stand well back while they zap my mouth with a tiny dose of X-rays. These days, doctors are aware that radiation has risks and that in the long term even a small amount of radiation raises the likelihood of cancer – by a small amount [🐕 170, 212].

When I was a child I remember going into a shoe shop with my mother to buy shoes, and looking into an X-ray machine that most good shoe shops had in the main sales area. By placing my feet just above a powerful source of X-rays, separated only by a thin sheet of aluminium, I could peer into an eyepiece and see the bones in my feet and the nails in the shoes I was trying on. There were eyepieces for the salesman and my mother so we could all enjoy the sight and ensure that I was not condemned to a lifetime of crippledom by badly fitting shoes. What I and other children might have been condemned to, of course, was radiation-induced sickness of various sorts, due to the doses of X-rays that were far higher than would be allowed nowadays for such a trivial purpose. In fact, these machines were no more than a sales gimmick at a time when science was universally good for you and the atomic age was all the rage, with its promises of free energy and nuclear-powered space travel.

The machines were promoted by advertisements like this one directed at parents:

> *They'll Need Their Feet All Through Life*
> Guard their foot health carefully through correctly fitted shoes. To help ensure a better fit, leading shoe stores use the ADRIAN X-ray Machine. Whether the shoe clerk is an 'old timer' with 20 or more years of fitting experience or a 'Saturday extra' who has been on the job only a few weeks, ADRIAN X-ray Machines help him give your child the most accurate fitting possible.

A radiologist writing recently about these machines described her horror at the leakiness of the radiation source: 'These X-rays not only went up through the feet to the fluoroscopy screen, but continued upwards, through the heads, thyroids and eyes of the users. Not exactly an optimal design. Even body parts not directly exposed to the main beam could receive a considerable dose from scatter radiation.' She showed a diagram with radiation contours stretching several feet from the machine and said: 'As an exercise in discomfort, look at the diagram and visualize just what sort of dose your own personal gonads would get while standing on this thing.'

It wasn't only the child users of these machines who could be bathed in X-rays. Salesmen working eight hours a day in proximity to this radiation source would have received much higher doses, particularly when they put their hands directly into the beam to squeeze the shoe during the fitting. Then there were the customers sitting in nearby chairs waiting to be fitted. The installation directions to shoe-shop proprietors said: 'We would suggest that you center the machine in the store so that it will be equally accessible from any point. Of course, it should face the ladies' and children's departments by virtue of the heavier sales in these departments.'

Once awareness of the dangers of radiation spread in the 1950s, as a result of the use of A-bombs in the Second World War and the radiation released into the environment by nuclear tests, some American states began to forbid the use for such a trivial purpose, and by 1963 they were banned all over the USA.

Because of the widespread use of the machines and the fact that radiation can lead to illnesses that have many other possible causes, it is difficult to be sure what harm they occasioned, apart from one reported case at the time of a model who had such severe radiation burns that her leg had to be amputated. But the subtler effects of radiation were likely to be long delayed and perhaps lost in the general figures for cancers that would have occurred years later. In 2002 and 2007, researchers reported on cases of a very unusual type of foot cancer which they ascribed to the use of shoe-shop X-ray machines forty or more years before.

STEMMING DISEASE

'Stem cells' is one of those buzzwords – or phrases – that get into the science news headlines, without people being quite aware of what they mean. On the one hand, these cells seem to offer hope for cures for all known diseases; on the other, they have become a political hot potato because of their association with moral and ethical issues to do with human embryos.

Setting aside their therapeutic or ethical significance, stem cells embody an extraordinarily sophisticated solution to a problem that puzzled biologists for many years. Everyone starts life as a single fertilized cell, which develops into a working body with hundreds of different cell types, merely by the process of cell division. That single cell becomes two, and the two become four and so on until a fully functional human being emerges, with the ability to eat and digest, move and grow, think and reproduce.

In the developed body, liver cells are very different from brain cells and stomach cells from blood cells, but they all start out as one cell type, known as stem cells – the stem of the 'plant' from which they all grow. The nucleus of a stem cell contains all the information needed to make any of the cell types the body will subsequently need. It is like a library with many different rooms, each with many sets of shelves. One room will contain manuals describing how to make and operate muscles; another set of shelves will give the blueprints for nerve cells and how to generate and transmit electrical impulses; a third will have all the instructions necessary to make a range of cells connected with the blood system, including immunity cells and red blood cells.

In the early stages of the developing embryo, the stem cells just reproduce. One 'library' spawns two others with identical sets of shelves and manuals. But at a certain stage, stem cells begin to differentiate into the more specialized cells that turn a block of identical cells into the beginnings of a body with a recognizable shape and functions.

When a stem cell differentiates, it is as if someone comes

into the library and locks all the doors to all the rooms except one, containing, for example, the instructions for making a liver cell. From then on, the library reproduces itself, along with the locked rooms, and for the rest of the organism's life, although it contains all the instructions for all body cell types, the descendants of that cell will have access only to the few open shelves, with the instructions for making a liver cell.

It sounds a cumbersome way of doing things. Why not just get rid of all the genes for other functions, so that future liver cells have only the information they need to do their job? One answer is that cells perform different functions at different stages in their lives, so they need to be able to unlock some of the locked rooms to consult the other manuals from time to time, particularly during the developmental phases of life, from infant to child, child to adolescent, adolescent to adult and so on. It also turns out that sometimes, later in life, cells that have become specialized still need to access the other information.

It may also be the case that it was actually simpler to evolve a standard method of cell division, copying everything, accompanied by other mechanisms for deciding later which genes to activate.

There are two ways in which scientists hope to use this phenomenon to treat or cure diseases. One way is to take healthy embryonic stem cells, with their potential to make any cell type, and find a way to implant them in people whose bodies may have defective cells of the sort that lead to chronic conditions like Parkinson's disease. This is caused by the presence of one type of defective nerve cell in the brain. If a stem cell could transform itself into a good copy of the defective cell type, the disease could be made to disappear. Stem cells can be treated to unlock only the 'rooms' containing the manuals for the necessary type of brain cell, and when they are implanted, they will provide the missing ingredient needed to cure the patient.

There is another type of stem cell, present in adults, that has had some but not all of its rooms locked, so that it knows how to manufacture new cells to replace those that 'wear out' or are destroyed by injury, such as heart cells, blood vessels, bone or cartilage. Since these also contain the instructions for making all other types of cells, scientists hope they may be able in the future to reprogramme them to unlock the closed sections and make other types of cells.

Finally, every cell in the body contains the instructions for making every other cell but – using the library analogy – the locks in the rooms are rusted shut, or jammed up with super-glue. Research to find ways of unlocking those instructions has succeeded in a small way by using cancer viruses to carry new genes into mouse cells which release the locked-up information. But this would be a dangerous technique to use in humans, because of the dangers associated with the viruses. However, in 2009 scientists announced the successful use of a technique to carry new genes into ordinary cells *without* using viruses in order to make the cells identical to embryonic stem cells, bypassing the use of embryos.

The way is open for new cell manipulations in humans that can be used to turn any cell type into any other, leading to a whole range of new therapies for intractable diseases.

RESISTANCE IS FUTILE

Medical advances are not always steps on a steady journey towards some ultimate state of perfect health. Some medical advances – such as the drug thalidomide or oxygen therapy for newborns [🐕 228] – *cause* problems as well as solve them. And the biggest medical advance of the twentieth century is leading to one of the biggest medical failures of the twenty-first, all because of the cleverness of bacteria in keeping one step ahead of the cleverest medical researchers.

The discovery of antibiotics transformed the way in which infections were treated. The two types of bacteria responsible for a significant proportion of all human infections – staphylococci and streptococci – were reined in almost overnight by the discovery and development of penicillin, followed shortly by other antibiotics. In the late 1940s and 1950s, sulfonamides, streptomycin, chloramphenicol and tetracycline, all more or less familiar names now, were developed to eliminate a whole range of disease-causing organisms.

But even in the early days, researchers began to notice that some strains of bacteria that should have succumbed to the new remedies survived and even thrived. Very soon after the drugs were put to use, it seemed that the organisms had been able to tailor resistance to the very drugs that had been designed to kill them.

Imagine if the advent of guns and bullets had led in a single generation to humans being immune to death by gunfire. Well, that's what was happening with some bacteria. As new drugs were developed to attack the resistant strains, bacteria acquired new types of resistance to the drugs. It became a race between the bacteria and the scientists, and by the end of the twentieth century the bacteria were winning.

Today, about 70 per cent of the bacteria that cause infections in hospitals are resistant to at least one of the commonest antibiotics. One of the most troublesome types of bacteria is a form of staphylococcus that resists several different antibiotics. It's called methicillin-resistant staphylococcus aureus or MRSA, and can sweep through hospitals, infecting many vulnerable patients.

But what does 'resistance' mean in this case, and how has it come about?

The ways that bacteria have found to defend themselves against the best efforts of drug manufacturers are a tribute to the ingenuity of evolution. Because bacteria reproduce so quickly, they show evolution in a very speeded-up form. In a laboratory, a bacterium like staphylococcus can divide and produce a new generation every couple of hours, compared with every thirty years or so with humans. If millions of these bacteria are exposed to a new drug, and just one bacterium acquires a mutation during cell division that gives it a degree of resistance to the drug, the descendants of that bacterium will flourish while the rest are killed, and before long that will be the dominant – or even the only – strain. Now the drug is useless.

But again, what does 'resistance' mean in this case? What does the mutation do?

There are at least three types of defence that a newly mutated bacterium can acquire. Let's go back to the 'immunity to bullets' analogy with humans.

If humans developed resistance to bullets in the way bacteria develop resistance to antibiotics, it could happen in at least three ways. First, we could develop some biological devices that roam around beneath the skin to trap a bullet as it arrives and snip it into many harmless pieces. Second, a similar device could swiftly turn the bullet round and send it back along the path it had travelled, 'bouncing' it out of the body. Third, we could develop a new type of skin that is impenetrable to bullets. These are analogous to three of the ways in which bacteria deal with antibiotics before they can cause them harm.

So far, the analogy holds – sort of. But bacteria have a way that humans don't of making sure that as many of their population as possible acquire this resistance, as quickly as possible. They can pass the abilities on to other members of the same generation, without waiting for their descendants to be born with them. For example, bacteria can actually pass on individual genes by physical contact with a neighbouring bacterium. Or they can leave bits of DNA in the surrounding environment which other bacteria can pick up. Or viruses that are specific to bacteria can carry the resistance genes from one bacterium and 'infect' another with them.

Once again, back to the analogy: it's as if people without immunity to bullets could acquire that immunity instantaneously through a kiss from an immune person, or by inhaling or being infected by snippets of DNA from such people floating around in the environment.

While it looks as if these amazing abilities have been called into action by the activities of medical researchers, there is a simpler explanation. It's likely that bacteria have always been able to create this kind of resistance, since certain types of bacteria produce their own antibiotics as a survival mechanism and they obviously need to be resistant to those antibiotics.

The spread of bacterial resistance to antibiotics seems inexorable. There are about one hundred million prescriptions for antibiotics every year in the USA, and many of them are for conditions like colds caused by viruses, which are unaffected by antibiotics. Nevertheless, the other bacteria in the body of the cold sufferer will acquire immunity to the antibiotic so that it will be useless later if the patient develops a bacterial infection. Even a simple action like stopping a course of antibiotics early can leave the surviving bacteria to multiply with their newly acquired resistance, again making future treatment more difficult.

ODDS AND SODS

THE SHIP THAT REPAIRED ITSELF

During the Second World War the largest ship afloat was the *Queen Mary*, 1,000 feet long and weighing 86,000 tons. There was a need at the time for aircraft carriers, preferably as large as possible to allow a variety of planes to take off and land, so when an English inventor, Geoffrey Pyke, came up with plans for a cheap and speedy way of building a ship that was twice as long as the *Queen Mary* and three times as wide, the Military Chief of Combined Operations, Louis Mountbatten, was very interested to listen to him.

Pyke was a shabbily dressed eccentric who had made and lost a fortune speculating in metals in the 1920s, and turned his mind to a wide range of inventions and innovations. In 1939, he sent a group of golfers on a tour of Nazi Germany, ostensibly to play the game but actually to seek the views of ordinary Germans about the Nazis, in the hope that Hitler would be shocked when he heard the true feelings of his people. But in August 1939 the golfers hung up their clubs, pocketed their balls and returned home with interesting data but too late to avert war.

Pyke's aircraft-carrier scheme used a material invented by a friend, Max Perutz (later to win the Nobel Prize for other researches), which Pyke modestly called Pykrete, perhaps an easier name to say than Perutzcrete. It was a simple mixture of 86 per cent ice and 14 per cent wood pulp, and Pyke showed that the addition of the wood pulp transformed the ice into a kind of 'super-ice' which could withstand bullets, resist crushing and easily be planed into blocks and other shapes.

The story goes that Mountbatten cornered the prime minister, Winston Churchill, in his bath, an occasional venue for important wartime meetings, and dropped a chunk of Pykrete into the bathwater to show its durability.

In spite of the odd nature of the material – and its inventor – Mountbatten set up a project to develop and test the material with the ultimate aim of building huge ships with the intriguing ability to repair themselves after torpedo attacks using on-board refrigeration equipment and seawater. Pyke even suggested that this equipment could be used to spray super-cooled water at enemy ships to ice their hatches shut and freeze their crews to death.

For a year or more the project progressed, with a top-secret laboratory in London and a prototype ship on Patricia Lake in Alberta, Canada. But the pace of the war and the technology of weaponry advanced to the stage where such potential behemoths as Pykrete aircraft carriers were superseded by longer-range aircraft, the destruction of the German U-boat fleet, and the rumoured development of a brand-new weapon, the atom bomb, that was expected to end the war.

After a post-war inventive fling, devising a fuel-saving system for railway freight wagons pulled by men on bicycles, Pyke became increasingly pessimistic about prospects for the world and about the resistance to some of his brilliant ideas and committed suicide in 1948.

WHIPS, THONGS AND CRACKS

If you were asked to make a list of 'the most misunderstood and misrepresented objects in our culture' you probably wouldn't put whips very high on the list. Furthermore, if you came across an article about whips in which the writer said he was 'indebted to A. Conway and P. Krehl for sharing their images of whips' you might think you'd picked up an issue of *Whips and Whipping* instead of a serious scientific publication. In fact, both these comments come in a scientific paper in the authoritative scientific journal *Physical Review Letters*.

A mathematician at the University of Arizona set out to investigate how a whip cracks. There had been reports dating back to the beginning of the twentieth century that the cracking sound was caused by the tip of the whip breaking the sound barrier. High-speed photography later in the century showed that part of the 'thong', the long flexible cord attached to the handle, undergoes an acceleration of 50,000 G. If experienced by a man, he would feel as if he weighed about three thousand tons.

Alain Goriely felt that the whip crack was probably some kind of sonic boom but wanted to understand in detail how the effect was produced. Previous explanations were no help. One scientist said that the narrowing of the thong meant there was a speeding up of the wave produced when the whipper moved the handle sharply downwards, until the very thin, almost hairlike ending was so light that it moved at the speed of sound. Another said that this couldn't be the reason because his calculations, using what's called linear momentum, showed that the tip of the thong should travel only at its initial velocity. Other scientists made calculations with different results, using another form of momentum, angular momentum, like the acceleration of a golf club during its swing.

'The purpose of this [article],' wrote Goriely and a colleague, 'is to reconsider the dynamics of the whip and reconcile these seemingly contradictory aspects such as the relationship between sonic boom and tip velocity, the effect of tapering, the boundary conditions, and the role of energy, linear momentum, and angular momentum.'

Littered with mathematical equations, the article roams over such topics as the classical Courant-Friedrichs-Lewy condition for stability, the motion of a movable accelerating elastic boundary in a supersonic flow, and the speed of sound in leather.

(A small diversion here – there isn't just one speed of sound. It varies with the medium carrying the sound waves. So in air, sound travels at 330 metres per second and in leather it's slower, at 220 metres per second. What's more, scientists have discovered recently that the speed of sound in moon rocks is much slower than the speed of sound in rocks on Earth, and suspiciously close to the speed of sound in cheese.)

The article by the Arizona scientists seems to settle once and for all the question of what happens when a whip is cracked. Where other scientists had calculated that it was the tip which travelled at just over the speed of sound in air and created the sonic boom, Goriely showed that the fastest part of the whip was a loop, created by the flick of the whip handler's hand, which travels at *twice* the speed of sound. The physics of this loop formation is apparently similar to the motions of the tail of a sperm as it swims towards the egg, although no one has ever measured tiny supersonic booms from that activity.

There's an old conundrum asking what part of a car travels at twice the speed of the car. The answer is the tops of the wheels. If the bottom of the wheel is stationary for a moment, as happens in normal forward motion, and the axle is moving forward at, say, 40 mph, then the top of the wheel will be going at 80 mph. In the same way, as a loop of the whip travels along from handle to tip, its top is travelling at twice the speed of the bottom. And because the thong is tapered, as the loop travels from a thicker to a thinner part of the thong, it speeds up, to as much as thirty times the initial velocity.

As often happens in science, a question answered leads to a new question. Could it be that the snapping of a wet towel is also the result of a sonic boom? In 1993, a group of schoolboys in North Carolina took high-speed photos of the tip of a snapped towel and showed that when the snapping sound occurred it was travelling at above the speed of sound. Setting aside the question of whether this was just an excuse to snap towels at each other, as schoolboys like to do, a suspicion was later raised that the object they 'snapped' was very unlike a real towel and could well have been constructed after their first snappings failed to go supersonic. There is a sentence in their report that says they 'made a new, longer towel from a piece of a cotton bed sheet'. This sounds dangerously like rigging the results. Perhaps it's time for Dr Goriely to come to the rescue. He's sure to know the speed of sound in towels.

Perhaps the most unusual insight to emerge from the investigation of supersonic whips is the conclusion that a certain type of dinosaur, the apatosaurus, probably used his tail as a whip in order to create sonic booms as a signal to other dinosaurs. The body of an apatosaurus was about one hundred feet long, with its tail taking up half that length. As anyone who looks at some of the larger dinosaur tails preserved in museums will have noticed, the bones in the tail get smaller and smaller towards the end, just like a whip. Computer simulations have shown that a wave travelling down such a tail could reach a speed of 1,300 miles an hour, fast enough to create a sonic boom as loud as a naval gun. This idea was confirmed by the fact that bones at the end of the tail, the part that would move fastest and suffer the most stress, were fused together, possibly as a result of the repeated stress of a sonic boom. A final intriguing part of the theory suggests that the dinosaur's sonic boom may have been used by males to attract a mate. About half of this type of dinosaur discovered to date have fused bones in their tails, and the recent discovery of a pair of dinosaurs, one male, one female, in Wyoming revealed that only the male had this characteristic.

WHAT'S THE DIFFERENCE BETWEEN A HYPOTHESIS AND A THEORY?

If you set up a row of dominoes on end from left to right and push the one on the left, it will fall on to the next one, knocking it over, and in turn this one will knock over the third one, and a ripple effect will eventually lead to the right-hand one falling over. The fairly obvious explanation for this is that pushing the first domino means that its centre of gravity is no longer above its base and in falling towards the second one, the pressure of contact will cause the second one to become off balance and so on. Anyone looking at this phenomenon will know, almost intuitively, why it occurs.

Now, imagine coming across two rows of dominoes set up in the same way on either side of a short cardboard 'tunnel'. Let's say you push the left-hand domino and then see the same ripple effect up to the tunnel, and after the sound of a few clicks from inside the tunnel you see the right-hand dominoes fall down in the same ordered way. How would you explain what you have seen? The chances are you would have little doubt that all that has happened is that a cardboard tunnel has been placed over the middle of the same row of dominoes as before, with the same cause and therefore the same effect.

Suppose, now, you come across a long cardboard tunnel on a table, with no sign of any dominoes. A red ball is rolled from the left into the tunnel and after a second or two a blue ball emerges from the right-hand end. Here, your search for an explanation is more open ended. You might – particularly if you have dominoes on your mind by now – have the idea that the red ball knocks down the first of a row of dominoes in the tunnel, and that the fall of the last domino pushes the blue ball out of the other end. But you might also think the red ball just ran into the blue ball in the middle of the tunnel and knocked it towards the other end (although why didn't it follow the blue ball out?). Or maybe the red ball ran into a white rat which had been trained to nudge the blue ball out of the other end of the tunnel. Or perhaps the red ball was actually some kind of hybrid of hedgehog and chameleon, with the abilities to roll itself into a ball and change colour. Or maybe. . .

What this has to do with is hypotheses and theories.

In the first example, the row of dominoes in clear view, it could be argued that no theories or hypotheses are involved. You *know* what is happening, from your knowledge of the physical world and the effects of solid objects hitting each other. If you had any doubts, you could do experiments on individual dominoes, weighing them, measuring the 'moment', a quantity that describes the force as it falls, seeing whether there were energy losses so that the effect might not be carried all the way down the line, and so on.

In the second example, with the dominoes either side of the tunnel, you have a *theory* for what is happening. You know how rows of dominoes behave when one is pushed, everything leading up to and away from the short stretch of tunnel is consistent with that explanation, and you have no reason to believe that there is not in the tunnel a similar stretch of dominoes to carry the motion from one side to the other. This theory is also by far the most likely explanation. In fact, there are no real candidates for another explanation. By using Occam's Razor, the principle that says the simplest explanation is likely to be the correct one, you believe that your theory is a full explanation of what you have seen.

In the third example, the red and blue balls, you have several *hypotheses*. Now, although there are no dominoes in sight, you might make an inference based on the possibility that there are dominoes inside the tunnel. But you are also open to other possibilities involving balls, rats or chameleon/hedgehogs. Among these competing hypotheses, there may be no way of knowing that one is as likely (or as unlikely) as any other.

This difference between a theory and a hypothesis is often misunderstood or even disregarded, in discussions of some scientific explanations of how the world works, for example. People sometimes say, 'Evolution by natural selection is only a theory,' in an attempt to cast doubt on it as a valid explanation for how life has evolved. In fact it is not *only* a theory, it *is* a theory, where little or no doubt is possible. It has the same sort of explanatory power as the theory that there is a continuous line of dominoes in the second example above. As with that example, there are gaps in what we know of the evolutionary path between primitive lifeforms and life on Earth today; as with that example, we know a certain amount about the left-hand end of the domino chain, from the study of fossils. We also know a lot about the right-hand end, from being able actually to observe and manipulate natural selection in the laboratory. As with the domino theory, anyone who doubts the explanation would have to produce some pretty convincing counter-arguments.

If people said 'evolution by natural selection is only a hypothesis', there would be room for alternative explanations, and perhaps intelligent design is one option (although I don't think so, since it lacks key characteristics of science). But evolution by natural selection is stronger than a hypothesis – it's a *theory*, the only theory that explains the facts and one of the greatest achievements of science over the last 200 years.

WAGON WHEELS

As the covered wagons race across the prairie pursued by whooping Indians, it sometimes seems as if their wheels are rotating in the wrong direction – at least if you are watching them in a movie. In real life this never happens.

This illusion is related to the fact that a moving picture is made up of a series of still frames projected twenty-four times a second. Each frame presents an image that persists in our visual system until it is replaced by the next frame [🐕 256]. If a film camera captures the movement of a wheel with spokes, such as on a covered wagon, the successive frames convey the rotation of the wheel by changes in the position of the spokes. There are in fact three possible illusions that this process can cause. The wheels could appear to stand still even though the wagon is racing along; the wheels can appear to go backwards; or they can appear to be going forward but much more slowly than you would expect. The explanation for the first example – stationary wheels, fast-moving wagon – makes it easier to understand the other two.

If we take a wheel with only four spokes to simplify the explanation, in frame 1 the four spokes might be in the form of a cross with vertical and horizontal arms pointing to twelve o'clock, three o'clock, six o'clock and nine o'clock. If the wheel rotates exactly a quarter turn between frame 1 and frame 2, the spokes appear not to have moved, because the one that was pointing to twelve o'clock is now pointing to three o'clock, and the others have all moved on by three hours. Because the spokes all look the same, we think that none of them has moved at all. If the wagon maintains its speed, the next frame shows the spokes having moved another quarter and so, again, it looks as if the wheel hasn't rotated at all. Over the few seconds the shot lasts, a hundred or more frames, it looks as if the wagon is sliding across the prairie on fixed wheels, provided it doesn't change its speed.

Now, supposing that instead of moving exactly a quarter rotation between frames, the wheel moves a little less than a quarter. Now, the twelve o'clock spoke moves to, say, two o'clock; the three o'clock to five o'clock and so on. In this case, the effect of the wheel moving forward by 'two hours' looks the same as the wheel moving backwards by 'one hour'. In the next frame, where the wheel is really moving forward by two more hours, it looks as if it's moved back by one 'hour', and the effect over the next few seconds is of the wheel slowly rotating backwards while the wagon speeds forwards.

With the third illusion, there's an appearance of the wheels rotating slowly forward as the wagon goes fast. This is because the twelve o'clock spoke has moved a little way *past* three o'clock, and so on, so that successive frames seem to show small advances of all spokes in a clockwise direction.

'SOME MUTE, INGLORIOUS DISNEY. . .'

In 1872, an eccentric English photographer in California, Eadweard Muybridge (born Edward Muggeridge), was asked to settle a question for a bet: when a horse is galloping, are all four feet ever off the ground at the same time? Muybridge developed a system using multiple cameras triggered by tripwires which took a sequence of still photos of a galloping horse, one of which showed all four feet off the ground. But Muybridge's system did much more than settle a bet. As a result of taking similar photographs of animals and humans in motion, he created sets of photographs which, when shown rapidly in sequence, produced some of the earliest moving pictures.

But according to some archaeologists, not *the* earliest. In the late 1970s, an Italian archaeological team discovered an ancient earthenware bowl in the ruins of a town in south-east Iran called Burnt City. This mysterious site, whose former inhabitants were farmers, builders and craftsmen, flourished 5,000 years ago and seems to have no connection with other civilizations in the area, in Mesopotamia, for example. Among the discoveries made over several seasons of excavation is a skull showing signs of some of the earliest brain surgery in the area.

When the earthenware bowl was discovered, archaeologists noted its interesting design, including a frieze around the outside edge, showing a pattern of drawings of goats alternating with trees. But no one considered that these might be frames in an animation, similar to the still images taken by Muybridge. It was only when an Iranian archaeologist, Mansur Sadjaji, copied each image separately and ran them in sequence that it became clear that the images were stages in the movement of a goat rearing up to eat the leaves off a tree.

No one knows, of course, whether this design was intended to be viewed in sequence. But an Iranian director has now made a film about the pot, which shows a convincing animation of the leaping goat. The effect is compelling and raises a couple of intriguing questions. True animation, like the motion picture, began in the nineteenth century with optical devices like the zoetrope, which could show moving drawings, and then the invention of cinematography. But these depended on a crucial insight into how the eye and the brain work. The zoetrope is a rotating cylinder with slits, through which you look at a strip of images, each slightly different from the previous one, around the inside of the cylinder. If you just rotate the cylinder and look at the images over the rim, the strip produces no illusion of movement, merely a blur. But when you look through the side of the cylinder, the moving slits produce the effect of a succession of still frames interrupted by brief instants of black. This is the same effect that is exploited in the movie projector, where a still frame is projected for a twenty-fifth of a second, then there is a moment of black before the next still frame. There are two puzzles about this process. One is, why don't we see a flicker, as the black and light images alternate? The second is, how does a succession of still images appear so convincingly as movement?

The absence of flicker is the result of a phenomenon called 'flicker fusion'. Even when a very brief image is flashed on to the retina, the information in that image is 'spread out' during the transmission from the eye to the brain. So the moment of black is swamped by the perception of the image if the interval is short enough.

But that doesn't explain why we see *movement* rather than a series of changing still images. It seems that there is further brain processing needed to interpret small changes in successive images as actual movement. There are even neurological conditions that affect this processing so that, in very rare cases, there are people who cannot identify objects in the real world when they are stationary but only when they are moving. Conversely there are others who have no trouble seeing objects at rest but the objects disappear when they move. One result of researches into these brain systems is the finding that, as far as motion perception is concerned, the brain activity when we watch films or animations is indistinguishable from what happens when we watch movement in the 3-D world, and that explains the extraordinary effectiveness of the illusion as well as the impact of cinematography when it was first shown publicly.

The discoveries and inventions of the last 200 years make it all the more surprising that a 5,000-year-old pot could be decorated with an animated cartoon strip. It would actually be theoretically possible to produce an effect of motion by putting such a pot on a turntable, rotating it and illuminating it using some primitive stroboscope. But this is a tall order. We can only assume that the artist was a very accurate observer of goat behaviour and hit upon the idea of representing the several stages of its motion as the animal rose to feed on a branch.

BEFORE BABEL

Is it possible to recreate the language of Stone Age man, spoken between 30,000 and 100,000 years ago? Some linguists believe it is. There is a small group of researchers who have traced words for key concepts in human life all the way back to a period when, they believe, the first language originated. They also believe that out of that first language, all 6,000 other languages on Earth emerged, from Abaza to Zuni.

Many of us are familiar with the idea that some languages show similarities in the words they contain and the grammatical structures they use. Italian and Spanish, French and Romanian, Russian and Serbo-Croat, Norwegian and Danish are examples. In the nineteenth century it became clear that there were wider and subtler connections, and languages were organized into families containing members that were not obviously similar at all – English and Sanskrit, for example, are members of the Indo-European family of languages. These 'family trees' were not organized on the basis of a hunch, but through the discovery of rules showing that words that appeared different in the way they were spelled or pronounced were actually derived from a common source. One of these rules, called Grimm's Law, explained how words beginning with 'f' in some European languages turned out to be derived from words beginning with 'p' in older languages like Greek and Latin. '*Pus*' (Greek) and '*pedis*' (Latin) led to *foot, fuss* and *fod* in English, German and Swedish. Similar changes occurred from b to p, g to k, and other sounds. Using these rules and comparing, say, English *mouse*, German *Maus*, Swedish *mus*, Russian *myš*, Polish *mysz* and Greek *mys*, linguists could 'reconstruct' the word for mouse in proto-Indo-European as – not very surprisingly – *mūs*.

But the Indo-European languages are just one group, with similarities that are often easy to observe, and they seem very different from the languages spoken in China, or among North American Indians. What if, as some linguists believe, there existed a Proto-World language, from which all human languages emerged? How would we reconstruct this?

In fact, this has been done for a small number of words, using similarities between language *families*, instead of between individual languages. One larger language family is called Nostratic, named after the Latin expression for the

Mediterranean, Mare Nostrum, around which some of the languages derived from it are now spoken. But the Nostratic
languages cover a much wider area today, and include a high
proportion of the language families of Europe, Asia, Africa and
North America.

By looking at words meaning 'seed', 'grain', 'wheat' or
'barley' in a range of modern languages, the linguists have
inferred that in the original Nostratic there was a word, 'bar' or
'ber', giving rise to *far* (Latin), *barley* (English), *burr* (Arabic),
paral (Malayalam) and a number of other words with similar
meanings in African and Indian languages, as well as
Mediterranean ones.

One of these linguists has written a poem in Nostratic, using
a specially devised alphabet to convey the early sounds:

> ḲelHä weṭei ʕaḴun kähla
> ḳaʎai palhʌ-ḳʌ na wetä
> śa da ʔa-ḳʌ ʔeja ʔälä
> ja-ḳo pele ṭuba wete

> *Language is a ford through the river of time,*
> *it leads us to the dwelling of the dead;*
> *but he cannot arrive there,*
> *who fears deep water.*

To many linguists, the approaches of these 'Nostraticists' are
seen as dubious. One of the critics has called them the 'Cosa
Nostratica' and accuses them of working in the 'Nostratisphere'.

But Joseph Greenberg, a pioneer in the field, believes
the techniques are rigorous. He compares the linguist who
tries to trace connections between modern languages with the
biologist who proves that a cat is closely related to a tiger and

more distantly related to a dog, and that both are descended from some earlier, now extinct ancestor – a proto-mammal – which can be reconstructed by reference to common characteristics of its descendants. For him, the next obvious step is to see whether the proto-languages can themselves be grouped into families, with the ultimate aim of tracing all languages back to one Proto-World language, as it has been called.

In fact, the theory that all languages are descended from a single starting point is not implausible, and is accepted by many linguists. But what the critics say is that that language is now too buried in the past to reconstruct or to find methods to prove that it existed.

To the proponents of the world language, the methods they use are rigorous and statistically valid. They are as sure that there were people who said *kuni* for woman or *wete* for water as I am that there are people who say *femme* and *eau*. Their arguments are often based on an impressive list of similar-sounding words discovered in many different languages, too many similarities to have occurred by chance, they say.

One linguist, Merritt Ruhlen, has no difficulty finding examples across the entire spectrum of world languages which suggest a common link. He gives as an example the word 'man' in English and finds the following examples across a wide range of languages. In the Bantu language Mbudikum-Bamum, *mani*; East Sudanic, *me'en*; Omotic, *mino*; Cushitic, *mn*; Avestan, *manus*; Gondi, *manja*; Indo-Pacific, *munan*; Amerind (Bella Coola), *man*; and Old Japanese, *mina*.

For a glimpse of what Palaeolithic *mano* or *kuna* might have been saying 30,000 years ago, here are some words of that first world language, as reconstructed by this new school of linguists:

aja	(mother)	aqwa	(water)
bunka	(bend)	bur	(dust)
kano	(arm)	kama	(hold)
kati	(bone)	kolo	(hole)
kun	(who?)	kuna	(woman)
mako	(child)	mana	(stay)
mano	(man)	min	(what?)
par	(fly)	poko	(arm)
teku	(leg or foot)	tika	(earth)

ESP NUCLEAR BLAST PREDICTOR

Many people believe in paranormal powers, in spite of the absence of evidence for such abilities as telepathy, clairvoyance and psychokinesis (e.g. spoon-bending). In the 1950s and 1960s, there was a burst of scientific interest in the field, and the hope that rigorous scientific testing of alleged psychics in laboratories would confirm the existence of such powers, and incorporate them into a scientific picture of the world. The laboratory showing most promise of a genuine effect was run by J. B. Rhine at Duke University, North Carolina.

Many telepathy and clairvoyance experiments used a set of cards with standardized designs so that the results of experiments in different laboratories would be comparable. Some individuals showed an apparent ability to 'see' cards as they were viewed in another room by the experimenter; others could tell what cards were going to be turned over at some time in the future.

In an important article in the American journal *Science* in 1955, an American geneticist, George Price, described a method by which precognition – the ability to 'see' the state of ESP cards in the future – might be used to predict future nuclear explosions. This was at a time when the cold war was at its height and many Americans feared an attack by the Soviet Union on American cities.

'Cards are prepared that will react to the thermal flash of a nuclear explosion,' Price wrote,

> so that the initial design will be bleached and a second design will develop. The cards are placed inside cameras with open shutters, surrounding a likely target area and directed upon various portions of the area. The cards are guarded and their symbols are kept secret. Each day several thousand selected percipients [people who claim ESP abilities] try to guess card symbols 10 days ahead. Guesses are analyzed in terms of each of the two possible correct symbols for each card.

Price suggested that the five standard ESP symbols be used and that each card have two symbols, one visible when it was in its unaffected state and the other revealed by the blast of an atomic bomb. If the cards were placed in vantage points around a target like Washington, DC, those that were pointing towards the site of an explosion, the Pentagon, say, would be bleached out and the new symbol revealed, while those pointing away, in the direction perhaps of the White House, would be unaffected.

So if one day the mass guesswork results in a statistically significant proportion of correct guesses of the hidden symbols on a set of cards in one area, while guesses for all the cards pointing away from the site of the future blast are significant for the usually visible symbols, the USA will have ten days' notice and the government can decide whether to evacuate the city – at the very least – or, perhaps more likely, to 'pre-retaliate' by bombing the likely aggressor.

'Does this suggestion seem absurd?' Price wrote, ironically.

> No. If information theory and Rhine's conclusions are both valid, this is a practical suggestion of high importance. Such a warning system would be far more effective and less expensive than radar. . . . [I]n general, any relationship between cards and guesses that is so highly improbable that it constitutes evidence for ESP can be made use of for transmission of information. And even if there is only 10-percent probability that Rhine's findings are valid, it is still the clear duty of appropriate government officials to investigate this possibility promptly and thoroughly.

The idea that the $14 billion cost of a programme like the US Strategic Defence Initiative, popularly known as 'Star Wars', could have been avoided by the purchase of, say, a hundred sets of ESP cards and a network of volunteer psychics is intriguing. As Price suspected, however, the work of Rhine and others was doomed to failure. Once cheating and lax experimental techniques were weeded out, no one was ever found who could guess cards above chance levels, although this didn't stop a university as prestigious as Princeton housing a parapsychology lab for thirty years until it, too, had to admit defeat and close down in 2007.

THE HAWTHORNE EFFECT

Most people have a broad idea of what a science experiment is. If you want to observe the effect of doing A to B, you make sure that nothing else changes while you do A, and watch B closely to note C, the outcome. If A is 'throwing in the air' and B is a coin, then C could turn out to be 'falling head up', 'falling tail up' or 'standing on its edge'. Normally, experiments are done many times to see whether the result C is repeatable. If the coin experiment is done once and the coin falls heads up, you would be rash to assume that this was always going to be the effect of tossing the coin. It may turn out to be the case after more tossings, in which case you have discovered a two-headed coin, but usually you will get roughly equal numbers of heads and tails and very few coins standing on their edge.

In a suburb of Chicago called Cicero, a shopping mall stands today on the site of a huge factory complex, the Hawthorne Works, covering 5 million square feet, where the conventional idea of how a scientific experiment works was overturned. More than forty thousand people worked on this site, making a wide range of telecommunications equipment, including the world's first high-vacuum tube, so it was a useful site for experimenters in the 1920s who wanted to study the effect of different working conditions on the productivity of workers.

One variable the researchers were interested in was the effect of different levels of lighting on productivity. So in a series of controlled trials, they changed the light levels in one area of the factory, from 24 to 46 to 70 foot-candles, but kept an adjacent area unchanged. Usually, scientists have a hypothesis when they do experiments, suggesting a range of results they anticipate but without a reason for opting for any one in particular. In the case of the Hawthorne experiments, there were three possible outcomes, as the researchers saw it – the better-illuminated workers could do better than their adjacent colleagues, they could do the same, or they could do worse. The result they got was none of these – the productivity of the experimental group increased with every *change* in lighting, either up or down. With the lighting at its lowest level, the productivity increased even more.

These results were not just confined to changes in the lighting, either. When the experimenters varied other aspects of factory life – maintaining clean workstations, clearing floors of obstacles, relocating workstations – every change in conditions led to increased productivity for short periods of time.

Over the years since the Hawthorne Effect, as it was called, was first described, there have been many further experiments producing similar results, leading to the conclusion that changing *any* variable in *any* direction usually increases productivity, even a change back to the starting condition. This might have been seen as a discovery with some practical use. Just run your factory with unpredictable light levels, random coffee breaks, changing scales of payment and variable working-day lengths and people will just keep increasing their productivity until they are working all day like whirling dervishes.

Of course, this turned out not to be the case. These increases, sometimes very dramatic, were not sustained. One review of recent work in the 1990s showed that a 30 per cent rise, typical of these experiments, often lasted only a few weeks and decayed to a small level after a couple of months.

It seems now that the experiments showed only that if people find themselves the focus of attention by management and researchers, with a degree of sympathy and interest on the part of people they respect or at least people who are their superiors, they will do better, perhaps without realizing it. Although this might not be a particularly valuable lesson for factory managers trying to get the most out of the workforce, it's an example of a scientific experiment that revealed something fundamental about how to do scientific experiments, at least where the subjects are human beings.

SOD'S CHAIN

Sod's Law, more politely called Murphy's Law, says that if something can go wrong it will. It's often believed to be responsible for such annoying events as toast falling butter-side down. A corollary to the law says that things don't just go wrong, they go wrong at the most annoying moment. Boilers breaking down in the depths of winter, the phone ringing as you're getting into the bath, and so on. While these things happen to everyone, and stick in the mind because of their annoying nature, they are usually interspersed with things not going wrong. If they weren't we'd never be able to go about our normal lives.

But occasionally, an example of Sod's Law comes up which is so far outside normal experience that we can be left in no doubt that, for a brief period, the normal laws of the everyday universe have gone on strike.

Such an example occurred some years ago and was described by a British pathologist in his collection of unusual clinical experiences. It was really a chain of Sod's Law events, each one of which would have been annoying, but for connoisseurs of examples of Sod's Law it has the extraordinary quality that if any single link in the chain had behaved in the way it was meant to, an ultimate tragedy would have been averted. But inexorably, as the story unfolded, the law operated on eleven occasions, each instance indicated by 'SL' below, and across two continents.

The story began with uncertainty over a pathological specimen that might or might not have shown evidence of cancer. Because the doctors in Hospital A could not agree, they decided to rush the sample to a pathologist in Hospital B, which happened to be in another country. (The particular condition was not always easy to discriminate and the other doctor was a specialist.)

Here is what transpired:

On Wednesday of Week 1, the envelope containing the sample tissues, on microscope slides, was taken to the post office near Hospital A, and put in the letterbox for airmail letters, addressed to Hospital B. On the label was written VERY URGENT – PATHOLOGICAL SLIDES FOR DIAGNOSIS – NOT DANGEROUS – NO COMMERCIAL VALUE. Unfortunately, the hospital clerk who posted the sample had not put quite enough stamps on it [SL1] and the post office sent it back to Hospital A, with a request for the correct postage to be attached. This arrived at the main reception, not the pathology department [SL2], in Hospital A on Friday, where it was decided that a hospital rule required the approval of a senior administrator to send an airmail package. (This rule was frequently circumvented by people paying out of their own pocket, as had happened initially.) A suitable administrator was not available until Monday [SL3], and he authorized the additional postage so that the package could be taken back to the post office and posted, which happened on Tuesday of Week 2.

After an uneventful journey of 12,000 miles (no, I don't know why there wasn't a nearer specialist either) the package arrived by special messenger at Hospital B on Thursday. Or, as it was called in that part of the world, Maundy Thursday [SL4]. Hospital B had a policy of central deliveries so the package didn't go to the pathology department but to a nearby office which, with the approach of the Easter weekend, had closed early [SL5]. The messenger handed the package to the only person remaining in the office, a cleaner [SL6], who put it quite appropriately in a box marked 'Incoming Mail'. (In fact, for the hospital's pathology department it was business as usual, dealing with the urgent tests that are generated every day in a busy hospital.)

The following Wednesday morning, Week 3, people returned from a relaxing Easter break and the package was handed to the pathologist. He looked at the slides and discussed them with his colleagues, and by four o'clock in the afternoon he came to a conclusion. He thought of telephoning Hospital A but realized that the time difference between the two countries meant that it was one o'clock on Wednesday morning at Hospital A. [SL7].

Later on Wednesday in Hospital A, the original doctor was worried that he had heard nothing from Hospital B, so he sent a telegram, asking for news. Unbeknown to him, the clerk who sent the telegram made a mistake, and addressed it to New York instead of the city and country where Hospital B was located [SL8]. Five days later, the telegraph company contacted Hospital A to report that the telegram had been returned as undeliverable.

Meanwhile, at Hospital B, knowing that the matter was urgent, the pathologist decided to send a telegram with his verdict, and wrote out the wording for his secretary. She in turn dictated it to Hospital B's switchboard operator and felt that she had done what was required. She was unaware, however [SL9], of a hospital regulation that required all telegrams to be approved by – yes, that's right – a senior hospital administrator. It would be disappointing for my story if at this stage things started to go right, but in fact the administrator in question was in a committee meeting [SL10] and then went straight home, so he didn't see the request for approval until late the following day, Thursday of Week 3. With a quick flourish he approved the telegram and on Friday morning his secretary took it to the switchboard operator, who telephoned it to the telegraph office.

With the speed of electricity – at last – the message in the telegram travelled to the reception area of Hospital A, where it was Friday evening. In a rare burst of efficiency, it was taken by hand *straight away* to the pathology department, where it was left in the office in-tray, unobserved until Monday morning of Week 4 [SL11], when the original pathologist opened it and read the opinion of his colleague on another continent.

In the nineteen intervening days, he and his colleagues had faced a dilemma. The sample *might* have proved the existence of a cancer in the patient's breast. On the other hand, it might have indicated that there was nothing to worry about. Uncertain what to do in the absence of any confirmation that the sample was benign, the medical team performed a mastectomy on the patient on the day before the telegram with the answer finally arrived.

The verdict of the pathologist at Hospital B, confirmed by the results of a biopsy on the amputated breast, was that there was no tumour.

If any *one* of the instances of Sod's Law had not occurred, the telegram would have arrived at least a day earlier, and the operation would not have been carried out.

I feel that such a sequence of mistakes one after another should perhaps be called 'Sod's Chain'. It may seem unusual, but if you look at some of the disasters of the last fifty years, a period when detailed inquiries began to be organized after particularly catastrophic events, such as Three Mile Island, Chernobyl and the two NASA space shuttle disasters, they all happened because of a Sod's Chain, and it was usually the case that if any one of the links in the chain had not been present the catastrophe wouldn't have occurred.

ANSWER THAT PHONE!

The earliest telephones, which went on sale in 1877, consisted of a wooden box with a hole containing a diaphragm. This vibrated in accordance with an electric current stimulated by the voice of the person on the other end. To hear the caller you had to put your ear to the hole, then to answer you had to turn your head and speak into it. The volume required to communicate with the new device was very high. As one observer wrote at the time: 'Telephone users held the receiver like a time bomb . . . shouted into the mouth piece at the top of their lungs[;] in fact, within six blocks, or ten if the wind was right, they could be heard without benefit of telephone at all.'

Initially, phones were marketed in pairs, to connect two locations – someone's office and his home, for example. But it soon became clear that the new invention would be much more useful, and more instruments would be sold, if it could be connected to a system of many users, so that any subscriber could talk to any other. But this primitive but effective device lacked one key feature that we now see as essential to any telephone system – a means of letting you know that someone was calling you.

If you happened to pick up the phone and someone was there, you could talk to them, but otherwise the device was somewhat limited. Initially, this difficulty was overcome by subscribers tapping on the diaphragm with a lead pencil, but the brittle material didn't survive very long under this treatment and had to be replaced. Alternatively, you could try shouting very loudly into the phone and hope that the person you wanted to speak to was near enough to hear the tiny squeaks coming from their box on the wall.

Then someone (not Alexander Graham Bell but a man called

J. C. Watson) invented a bell that could be built into the apparatus and made to ring when someone made a call. This was a distinct improvement, apart from one drawback – there was no way to make an individual phone ring. In one of the earliest telephone systems, in Toronto, to make a call, a subscriber had first to speak to the operator, who would then ring all telephones on that system. Each person would go to the phone to discover whether the call was for him or not, and if not, common decency suggested he should step away, although there was nothing to stop him staying near by to listen if the call sounded potentially interesting.

Faced with a situation in which every time anyone made a call, all the phones in the town rang, T. D. Lockwood, author of *Practical Information for Telephonists*, wrote that 'the constant ringing of bells, melodious as it might be per se, and sweet as the Bells of Corneville [a popular musical of the time], yet became a trifle monotonous, and the wearied ear yearned for rest and silence'. Eventually, a solution was found that allowed the operator to direct calls towards the correct subscriber, although there were still plenty of subscribers who had what were called 'party lines', which meant that they had to share with a few other subscribers.

What was needed was a device that allowed people to direct their calls without having to go through an operator. The device that allowed this to happen came about because its inventor, an undertaker called Almon Strowger, didn't trust the local operator to put calls though to him. She was the wife of another undertaker and he suspected her of diverting calls from newly bereaved relatives to her husband when they might have been intended for him.

His invention, the Strowger switch, required subscribers to tap out the digits of the number they required with a button, and this would operate a contact arm at the exchange which could move into any one of a hundred different positions, enough for the small number of people who had phones in the exchange area.

Strowger became a rich man, but in 1902 his undertaking firm received a phone call from his wife, to say that Strowger had died and required their services.

HOW MANY PIANO TUNERS IN CHICAGO?

The physicist Enrico Fermi was fond of asking this question in his lectures to students in Chicago, to show how most people can come up with answers to questions about which they have no expert knowledge, if they make rough guesses based on ordinary assumptions.

He showed how a whole range of scientific questions could be answered with very little basic knowledge – guesswork really – along with one simple rule. With his piano-tuner question, he'd show how the roughest guesses about population size, the proportion of people with pianos, how often pianos require tuning and so on, can be used to arrive at an approximate answer. He then showed students that they could use similar guesswork to arrive at useful answers to questions about science, such as: What is the mass of the Earth? Who walks faster – short people or tall people? How much food goes into physical work and how much into keeping you alive? What's the total mass of all the students in your school or co-workers in your office? How many cells are there in the human body? These questions have become known as 'Fermi questions', and they are often used in teaching.

To take a question that doesn't even require knowledge of Chicago or pianos, how many years of life does the average smoker lose? Well, you may complain, 'in place of Chicago and pianos I now have to know about cancer and life expectancy and stuff like that'. But most of us absorb all sorts of things from reading the newspapers which enable us to make educated guesses.

Supposing you were forced to answer this question, with a gun to your head or even just pincers to your fingernails – how would you try to save your life, or your nails?

You probably know that smoking kills largely through cancer and heart disease and that these are diseases that arise mainly in the over-fifties. You also know that most people don't live beyond eighty. So the answer must be somewhere between nought and thirty. It would be nought if no smoker's life was ever shortened by the habit, and thirty if every smoker died on his or her fiftieth birthday. Now clearly it's not nought, otherwise there wouldn't be a problem and we wouldn't have asked the question. So let's say the answer is at least one year and could be as much as thirty years.

So you have what mathematicians call an upper and a lower bound [🐕 71], thirty and one. Now the simple rule I mentioned above comes into play: you have to work out what is called the 'geometric mean' of your upper and lower bounds. You get this by multiplying the upper and lower bounds together and taking the square root. 1 × 30 is 30, and the square root of 30 is just over 5 (5 × 5 is 25, 6 × 6 is 36).

With these simple pieces of knowledge – or even just guesswork – we have arrived at some kind of answer – 5. The actual answer is 6.5 so we've not done too badly.

There's one other tip for dealing with Fermi questions. Pi can be 3, days have twenty-five hours, every adult weighs about 10 stone, spheres of diameter d have roughly the same volume as a cube with side d, and so on. Such approximations are useful enough and sometimes the effects even cancel out.

Getting an answer as close to the correct one as 5 to 6.5 is not a fluke. Using these techniques will usually produce a number with what scientists call the right order of magnitude. This means that your answer will probably be out by no more than a factor of ten. This may not seem very useful and in real life it isn't. You'd like to know your salary, your pension, your weight or your disease prognosis with considerably more accuracy, but sometimes in science any answer is better than no answer at all.

You can use the same rough-and-ready techniques to answer questions like 'How many people are airborne over the USA at any given moment?' or 'How long a hot dog can be made from a typical cow?' and get an answer that is within the right ball-park. (You could even work out how many planes or hot dogs would fit into the average ballpark.)

ARAB SCIENCE UNDER THE TELESCOPE

The word 'chauvinism' is often used to mean 'male chauvinism', the belief that men are superior to women. But its correct meaning, fanatical patriotism, is derived from a semi-mythical character in several French plays in the early nineteenth century, Nicolas Chauvin, who was an excessively nationalistic soldier serving in the army of Napoleon Bonaparte.

A modern example of chauvinism, which actually has a grain of truth behind it, is an essay I came across in a Libyan manual of the English language, by one Mohammed A. Manna', which lays out the Arab contribution to science in order to set straight the historical record. It is called *The Arab Scholars*, and I give it with its original spellings and grammar:

It is regretted that the western writers and the orientalists in particular have always passed over without notice the works of the Arab scholars.

The European writers have occassionaly fabricated defamatory stories with misrepresented facts over the Arabs ability to have taken parts in any exceptional skill due to knowledge.

In order to illuminate to a great extent the way, it is therefore a strenous duty to an abbreviate hints over certain works rendered by Arab scholars so as to frustrate false allegations.

There is no doubt that whilst Europe was immersed in the darkness of ignorance and whilst America was unknown to history, the Arabs had an outstanding figure on the stages of science.

The universal number had been unknown to Europe until Jaber El-Ash'bily of 1196 A.D. made Algebra which is nowadays still holding his name.

In astronomy, lbn Elheitham Albasary changed the course of history by contriving the microscope that made astronomy a practical real event. It is understood that England's Beacon invented the modern microscope immediately after he had the books of Albasary in which he described the space being dark and perpetual night, and that only in the atmosphere surrounding the planets is there light the reflected light of the sun.

In medicine, the Arab scholars were the first to excel in anatomy and surgery and they also played a considerable roll to achieve pharmaceutic drugs and heal several infectious diseases such as smallpox, measles, scab, dysentry, dysury, miscarriage, plague, cholera and others; and diagnose all sorts of pathogenesis.

Whilst the Europeans were practising jugglery as means to cure the patients deceiving them by trickery and while they believed that the diseases had been an 'evil' invoked by nature, the Arab scholars considered the theory as a retrograded worthless to trouble and they found out 'contagion' which is the communication of disease by contact with a person suffering from it, before Louis Pasteur came into being.

Moreover, the Arab scholars versed in Chemistry and performed the process of analysis in addition to their contribution relative to inorganic Chemistry which deals with mineral substances; organic Chemistry with animal and vegetable substances and stepped forward to intermix Chemistry in fine arts such as tanning, wool-dying with assorted colours, tinning and cosmetic composites.

In addition to such scientific conquests they – in spite of being nomads – had not connived at agriculture in which the books of Zakariya Al'ashbeily evidently denote the Arab far-reaching effects on cultivation the soil whether by posturage, farming, tillage or horticulture, and the management of land and also the study of plant nutrition, irrigation systems and the convenience of climate to a certain soil for a definite plantation period.

By the end of 10th Century the Arab scholars invented the materials of distillation, stocks of filtration and preparation of alcohol and the operation of metallurgy.

No modern scientist dare deny that the words of alcohol, alembic and potash still keep their Arabic origin.

The music was a derision, but it flourished ten centuries ago when El-Faraby composed the first musical note on whose symphony the Spaniard teenagers are so far dancing.

There had been no geography at all until El-Idrissy outlined the first map of the world displaying the physical features of the Earth's surface and the distribution of land and water; continents and oceans. Christopher Columbus, the famous navigator, discovered West Indies five centuries ago immediately after he had acquainted El-Idrissy's charts on which he relied during his voyages.

The Arabs founded manufactures of iron and steel; weapons and ammunitions; glass and crystals; tannaries and factories of textiles and silk, weaving wool, cotton and linen at Iberian Peninsula.

Furthermore, Abdurrahman El-Badr was the first to invent lithography and typography eight centuries before the German Goettenberg came to light.

The Arab scholars have over and above shown a great aptitude in geometry, metaphysic and philosophy, Muhiddin El-Araby and El-Maarry left a great literal production from which the modern philosophers cited too much. It can never be hidden that the 'Description of Paradise' of Italy's Dante Alleghieri was quoted word from 'Risalat El-Ghoufran' of El-Maarry.

Ibn Khaldoun wrote his hieratical and sociological books before British Samuel, Shakespeare and Becon, French Descartes and German Arthur Schopenhauer came to light.

It is evident that whilst the other nations were living in

caves and huts, the Arabs architectures appreciating beauty in all forms constructed and adorned palaces, houses and mosques throughout Southern Europe, Africa stretching to the Far East.

The orientalists have cited several scriptures of Chemistry and mechanism from the books of Ibn Sina (Avecenna), El-Ghazaly, Ibn Erroumi, Errazi and others. Not only have the orientalists appropriated the works, writings, inventions and ideas, theories of the Arab scholars copying them in to make over the possession of, but they denied the presence of the real compilers.

It is too difficult to gather the works of the Arab scholars in an abbreviate article as it requires too many volumes to contain.

It is not essential to demonstrate pompously the great works of our progenitors but owing to permenant illimitable groundless aspects wrongfully transpired by certain orientalist and extremist elements financed by zionists, it is therefore inevitable to stultify false allegations being asserted without proof, and on the other hand display the works of those genius scientists who have paved the way to 20th Century Civilisation.

When life gets me down, as an Englishman of Arab origin (indeed, my surname is derived from the craft of 'wool-dying with assorted colours'), I get great pleasure from reading *The Arab Scholars*, with a glass of alcohol by my side, and the first musical note composed by El-Faraby playing on the hi-fi. And yet, for all his occasional impenetrability, unconscious humour, solecisms and shaky command of English, Mr Manna' has presented a roll-call of philosophers and scholars, most of whom are unfamiliar to the Western reader, who made significant and often fundamental discoveries which underlie much of modern science. But surely Manna' missed a trick by failing to point out that Dante was probably Dante Ali Ghierri. . .?

WHY DOES A MIRROR REVERSE LEFT AND RIGHT BUT NOT TOP AND BOTTOM?

This is a *very* interesting question, not because it is particularly interesting in itself, but because it produces so many silly answers. In fact, I've produced quite a lot myself. Here are some extracts from my attempts to tackle it:

'Let's start with one basic fact – mirrors give no preference to what they reflect. . .'

'It turns out that left and right are not absolute definitions, like up and down. There is no left or right side to a pyramid, for example, although there is a top and a bottom. . .'

'"Left" and "right" are not so easily defined, and you'd have a much harder time telling a Martian what you mean by these. . .'

'. . .because you normally place a mirror vertically. If you put a mirror on the floor, it reverses up and down just fine. . .'

'Suppose there was no gravity. Then we may meet people oriented 180 degrees along the x-axis (where the x-axis is defined as from left to right, straight through their right and left sides). Thus, their feet are where our head is, and their head is where our feet are. . .'

Now that I have the real answer (see below), at least for today, these early attempts seem to be groping in the right direction but not really hitting the spot.

Richard Dawkins found that this question provided a useful way to get university candidates talking:

> For years as a college tutor at Oxford, I would try the intelligence and reasoning powers of entrance candidates by asking them at interview to muse aloud on the conundrum of why mirror images appear left–right reversed but not upside down. It is a provocative puzzle, which is hard to situate among academic disciplines. Is it a question in psychology, in physics, in philosophy, in geometry, or just commonsense? I wasn't necessarily expecting my candidates to 'know the right answer'. I wanted to hear them think aloud, wanted to see if the question piqued their interest and their curiosity. If it did, they would probably be fun to teach.*

Well, Professor Dawkins, let's see whether I would be fun to teach.

The first thing I'd like to do is set the mirror aside for a moment – it just confuses the issue. Here's a related question – why does a right-hand-glove fit your left hand when you turn it inside out, whereas when you turn a hat inside out it doesn't become a pair of shoes? The transformation that turns a right hand into a left hand is known mathematically as a reflection, whether or not a mirror is present. Many properties do not change during that process – the colour of the glove, the lengths of the fingers, the angles between the lines of the pattern on the material, and so on. All of these characteristics – the lengths, the angles, the patterns, the handedness – are properties of the

---

*Richard Dawkins (ed.), *The Oxford Book of Modern Science Writing*, Oxford University Press, Oxford, 2008.

glove, some of which are affected by the mathematical process of reflection, others not. But nowhere in this glove is there a fixed up or down. On a raised hand, its fingers might be up and its sleeve down. In space, a hundred light years away, it would still be a right-hand glove but nowhere would be up, which relates only to objects on Earth.

So the words 'right' and 'left' in the question at the top of this essay are fundamentally different from the words 'top' and 'bottom'. I suppose what people are thinking of is 'head' and 'feet', and they wonder why the head and the feet are not flipped too, when the left–right change occurs.

But when you think about it, we *define* our left and right sides with reference to our head and feet. Your right side is made up of the first parts of your body you come to as you imagine a clockwise path around your body – head, right side, feet, left side, head. But in a gravity-free environment we might have oriented ourselves by defining one hand as left and the other right, and then the top of your body, including your head, would be defined *relative* to this handedness. You could define top – where your head happens to be – as the first parts of your body you come to as you imagine a clockwise path starting at your left hand. In that case, the person with the body that is reflected in the mirror would find that the part of his body that should be defined as the top, or head, i.e. the first part of the body that was reached in a clockwise direction from his left hand, would be the part wearing the shoes, and he would be asking, 'Why does the mirror reverse top and bottom but not left and right?'

So the answer to the initial question is another question: Why should it?

I think I feel better now.

COMMUNICATING AT AN UNKNOWN RATE

Two men – let's call them Bouvard and Pecuchet – stand on two hills in France. One hill is the site of a town under siege, the other, about six leagues (eighteen miles) away, is the source of possible relief. Each man has a small cannon by his side and holds a pendulum in his hand. The man under siege, Bouvard, fires his cannon; Pecuchet on the other hill fires his cannon back and starts his pendulum swinging. When Bouvard hears the sound of Pecuchet's cannon he lets *his* pendulum go and it starts swinging. After a certain number of swings – let's say twelve – Bouvard fires his cannon again. When Pecuchet hears this, he stops counting the swings of *his* pendulum and opens a small booklet. It is a list of words and phrases associated with numbers. Next to number 12, the phrase is 'I need food'. Pecuchet shouts to his servant to organize a mule train laden with baguettes, strings of onions and quiches Lorraine, which sets off for the besieged town. Several hours later, Bouvard fires his cannon again, the pendulums swing twenty-three times and Pecuchet looks up his booklet to find, at number 23, the phrase *Merci beaucoup*. The besieged citizens have been saved from starvation.

In a world in which almost instant communication, such as telephone calls, Internet traffic or television transmissions, is the lifeblood of society, it's difficult to imagine how recently fast communication was invented. It's also surprising to learn how slow and cumbersome communication across distances was only 200 years ago.

The pendulum method of communicating at speed over long distances was proposed in 1790, in the *Encyclopedie Méthodique, Arts et Métiers Mécaniques*, although there is no evidence it was ever used. Over the centuries up to this time, a range of methods had been used fitfully to fulfil a growing need for communication over distances which was faster than a man on a horse. The Palestinian poet Mikhail Sabbagh (yes, a distant forebear of mine) wrote a definitive text on carrier pigeons, including details of how they had been used in historical times, starting with Noah, the dove and the olive branch, and going on to postal systems operated under various Muslim caliphs.

At the time of the French Revolution, the importance of government and military communication grew to such an extent that a law was passed by the French parliament that said: 'Anyone who transmits any signals without authorization from one point to another whether with the aid of mechanical telegraphs or any other means will be subject to imprisonment for a duration of between one month and one year. . .'

No one underestimated the value to government and commerce of a successful system of communicating from one end of the country to the other and, indeed, between countries. Unfortunately for the inventor of 'pendulotelegraphy', within a few years of the system's description, a 'mechanical telegraph' had been invented by the Chappe brothers. On 2 March 1791 at 11 a.m., they sent the message *'si vous réussissez, vous serez bientôt couvert de gloire'* (if you succeed, you will soon bask in glory) over a distance of 10 miles, using a combination of black and white panels, clocks, telescopes and code books.

The system was so successful that it came to operate over a 5,000-kilometre network in France linked by 534 stations. To transmit one word between stations, using a contraption with movable arms at the top of a tower controlled by cables operated below, took about three minutes, and for a message to travel from, say, the port of Toulon to Paris, a distance of 800 kilometres, it would pass through eighty or so stations, and take several hours. In Alexandre Dumas' *The Count of Monte Cristo,* a financier's downfall is caused by bribing a telegraph operator to change the content of a message on its way from the south of France to a government department in Paris.

But such systems of telegraph towers, including one organized by the Admiralty in Great Britain to bring messages from south and east coast ports to London, disappeared almost overnight with the invention of the electric telegraph, which came into commercial use in the 1840s. The only trace of the clunking mechanical telegraph today is a smattering of place and street names, such as Telegraph Hill, on the maps of England and France (and San Francisco).

SWINGING THE LEAD

When I was a child, Christmas crackers used to contain a mysterious and amazing device which puzzled me for a long time. It was a simple pendulum, a piece of thread and a 'bob' consisting of a small piece of brass-coloured metal stamped out in the shape of an arrow or a fleur-de-lis. If you held this over the hand of a boy or man and waited a short while the bob would move backwards and forwards in a straight line; if you held it over a girl or woman, it would move in a circle. It even indicated the sex of the family pet. Furthermore – even more amazing – if you held it over a pregnant woman, the motion would indicate the sex of her unborn child.

I can't say I believed it 100 per cent. Being a sceptical little boy, I was more interested in understanding what was going on, and that meant also understanding how it could be explained. Preferably this should be done without expanding the boundaries of scientific knowledge to include some kind of sex-linked emanation from living creatures that affected the movement of a pendulum.

But I have to say, that 'pregnant woman' thing made it very difficult to accept one possible explanation – that since you know the sex of the creature you are holding the pendulum over, you somehow – consciously or unconsciously – move the pendulum yourself in the desired direction. How could that explain a definite movement in a circle over the tummy of a woman who did not know the sex of her unborn child and later had a baby girl, for example?

Now that I am older and a bit wiser, I know that I was right to be cautious. There is nothing mysterious about what is going on, and the amazement relates only to the ability of the human body to use visual feedback in a subtle and interesting way.

What is actually happening is the following:

First, the effect *does* depend on knowing the sex of the person or creature (we'll come to the pregnant woman later).

Second, the bob will move the way you expect it to – if you are told the straight line means 'female' and the circle 'male', that's what will happen. (If you are told the opposite, that will happen instead.)

As you hold the pendulum steady and 'will' it to move, random tiny muscle movements in your fingers begin to transmit movement to the thread. Some of those movements will cause the end of the pendulum to move very slightly in the appropriate direction, and as long as you can see the way the pendulum is moving, those particular muscle movements will be increased, whereas the ones that caused the opposite type of movement will be dropped. As you watch closely and the pendulum bob moves more and more in the appropriate manner, your fingertip muscles will become more effective at transmitting precisely the moves that make the effect happen.

There's an easy way to check whether this is the explanation. Try willing the bob to move in one or other of the patterns, without a hand or person underneath. You can make it do so at will, however steadily you hold your hand. Try to switch from straight line to circle for a man and the effect will occur. Try to 'will' the pendulum to rotate (without deliberately moving your hand) but keep your eyes closed. Nothing will happen, confirming that visual feedback is necessary so that the correct muscle movements are 'rewarded' and enhanced by a visual indication of success.

And here's one final dramatic demonstration of the power of tiny muscle movements to respond to visual feedback. Hang the pendulum in a bottle, pinching the top of the thread between the cork and the bottle neck, and put the bottle on an uncovered tabletop. Watch the bob carefully, with your hands pressed flat on the table. Now 'will' it to move in a circle or a straight line. It will take a bit longer than when you are actually holding it, but sooner or later the bob will obey your instructions, as a result of the tiny muscle vibrations being transmitted through the wood, into the bottle and on to the thread.

And the pregnant woman? Well, I have to confess that the bob *does* seem to predict the sex of the baby, but only 50 per cent of the time. . . A tossed coin will do that, too.

EUROPE TO AMERICA IN AN HOUR, BY TRAIN

When an American lawyer, Frank Davidson, and his young family spent seven hours tossing in a ferry in the English

Channel in 1956, he decided that there must be a better way to travel from France to England. Because he was well connected in the American financial world, Davidson told the story of his experiences over lunch to some New York banker friends, and in no time they had set up the Channel Tunnel Study Group. A mere forty years later, as a result of that initiative, trains in the Channel Tunnel were carrying cars and passengers between the UK and France, and seasickness was no longer compulsory for people who didn't like flying.

Davidson is still active, in his eighties, and for the last two decades, with a small group of engineers at the Massachusetts Institute of Technology, he has been advocating a transatlantic rail line, running in a tube under the ocean from, say, Bristol to Boston, using what are called maglev trains. The technology involved – magnetic levitation – uses the repulsion between magnets on train and rail to create a frictionless cushion for the train to ride on, and there is a test track in Japan showing that the system can reach speeds of 300 mph.

But who would want to set off on a 3,000-mile train journey just to get from Britain to America? Even at 300 mph, the journey would take ten hours, compared with five or six by air. Frank Davidson's 'Atlantic Tube' would, however, have one extra characteristic: there would be no air in the tube so the train would run in a vacuum, saving the energy wasted through friction with the air in a conventional maglev system and making it possible for the trains to travel much faster.

In this way, Davidson's train would beat the plane by a comfortable margin. The technology he advocates means that there would be no limit to the speeds that could be achieved, since the resistance would not increase with the speed, as it does in air. Trains could easily travel at speeds of 5,000 miles an hour and more so that a day trip to the USA would be possible, with the journey taking an hour each way.

Unlike many predictions of future technology, this system will not depend on new ideas. There are only two obstacles – finance and public acceptability. To construct the tube, a simple process of manufacturing concrete sections in countries around the North Atlantic and assembling them simultaneously, is not a difficult task. The propulsion technology exists already. And increasingly sophisticated GPS satellite technology could overcome perhaps the trickiest problem of all – how to prevent accidental contact with the walls of the tube at high speed. (The friction would incinerate train and passengers.)

The Atlantic Tube would require a major collaboration over funding and construction between several countries and international corporations. But the eventual cost to passengers would be surprisingly affordable. When estimated a few years ago, likely ticket costs were about £100 each way, very competitive with air travel and much quicker.

FUNDAMENTAL CHEMISTRY

Heterocyclic compounds are molecules made up of atoms arranged in a ring, usually of carbon and hydrogen, along with another atom of a different element. There are many different heterocyclic compounds, with different numbers of atoms making up the ring structure and a variety of elements supplying the other atom. They all have unique chemical formulae, such as C_5H_5N or $C_4H_8O_2$, but it's often more convenient to refer to molecules by name. There are many existing variations of this particular type of molecule as well as the possibility of synthesizing new ones, and in the 1880s two chemists, named Arthur Hantzsch and Oskar Widman, independently came up with a system for determining a unique name for each compound based on the number of atoms in the ring and the nature of the other element that is present.

Since this is not a chemistry textbook, you may suspect that this essay is heading somewhere less arid than the rarefied heights of chemical nomenclature. And you'd be right. The Hantszch–Widman nomenclature, as it was called, failed to take account of the puerile sense of humour of British chemists. For almost every heterocyclic compound, the system produced a perfectly unexceptionable name that was memorable without being indecent. But a problem arises when the element supplying the extra atom is arsenic. This word is derived from an Arabic word, *al-zarnikh*, and Professors Hantzsch and Widman decided that all heterocyclic compounds using arsenic should begin with 'ars'. (Other elements gave rise to prefixes like 'fluor' for fluorine and 'iod' for iodine and so on.) So far so good. But they then decided that, depending on the ring size, there should be a series of suffixes, such as 'irine', 'ete', 'ole', 'inine' and 'epine'. So unfortunately there was no way to avoid a compound based on an atom of arsenic plus a size-5 ring being called 'arsole'.

This is a genuine molecule. You can read more about it in a paper by two Swedish scientists which begins: 'The aromaticity of arsoles has been debated in the literature for years. . .' And Paul May, a chemist at Bristol University, is so taken with the possibilities of this nomenclature that he suggests on his website 'Molecules with Silly or Unusual Names' that the synthesis of arsole and a 6-benzene ring would be called 'sexibenzarsole'.

ACKNOWLEDGEMENTS

The following people have been helpful in various ways during the writing of this book, either by suggesting topics – sometimes without being aware of it – or reading drafts, and I am grateful for their assistance:

Iain Chalmers, David Childs, Frank Davidson, Marc Dion, Christy Hawkesworth, Nicholas Humphrey, Harvey Marcovitch, Richard Newell, Martin Rees, Linda Smith, Bill Silverman, Max Walters, and Larry Weiskrantz.

I would also like to thank John Brockman for permission to include a revised version of an essay I wrote for *What Is Your Dangerous Idea?*, the book which he edited, published by Harper Perennial in 2007.

Finally, I am grateful to Roland Philipps at John Murray for his encouragement and advice during the writing of this book.

Notes on sources can be found at:
www.sabbaghhairofthedog.wordpress.com